Technology and International Affairs

Technology and International Affairs

Edited by

Joseph S. Szyliowicz

PRAEGER SPECIAL STUDIES • PRAEGER SCIENTIFIC

Library of Congress Cataloging in Publication Data

Szyliowicz, Joseph S.
  Technology and international affairs.

  Includes index.
  1. Technology and international affairs.   I. Szyliowicz, Joseph S.   II. Title.
T49.5.S98   303.4′83   81-13985
ISBN 0-03-053321-X   AACR2

Published in 1981 by Praeger Publishers
CBS Educational and Professional Publishing
A Division of CBS, Inc.
521 Fifth Avenue, New York, New York 10175 U.S.A.

© 1981 by Joseph S. Szyliowicz

123456789   145   987654321

Printed in the United States of America

To Irene, Michael, and Dara

Preface

The past 25 years have witnessed a remarkable proliferation of academic programs that are concerned with the relationship between science, technology, and society. These programs have emerged as a response to the obvious fact that we live in an age when the influence of science and technology cannot be ignored. Modern technologies have permeated every part of the globe and have provided man with an unprecedented capability to alter human life, cultures, societies, and the global environment as well. Indeed, technological change has become so rapid that we are confronted with both potential benefits and dangers on an unprecedented scale. Changes have already occurred and continue to occur in people's lives everywhere, in the role and power of nation-states, in the structure and functioning of the international system itself. These changes have seldom been anticipated or planned for, even within the advanced societies.

Concern with the ways in which technology impacts upon human affairs has been accompanied by a great increase in scholarly activity in this area, and we have witnessed a remarkable outpouring of applied and theoretical works that seek to analyze the ways in which technology impacts upon societies, local, national, and international, upon the dynamics of the technological process, and upon the development of tools and techniques that would facilitate the production, assessment, and control of technology.

Such work originated with scientists and technologists who were naturally most aware of the power of the new technologies but, increasingly, social scientists have become active in this area as well, as it became ever more evident that the critical variables involved in the utilization and control of technology were political, social, and economic, as well as technological.

Yet, sophisticated analyses of the interplay between technology and international affairs remain relatively scarce, and many persons involved in college and university programs in this area have felt the need for additional teaching resources. Accordingly, when I was invited by the International Studies Association to organize a series of panels dealing with this topic for the 1978 convention, I accepted the responsibility because it gave me an opportunity to bring together knowledgeable scholars and practitioners before an association whose members had, by and large, paid little attention to the role of this important variable. Accordingly, I organized three panels. The one that I chaired focused on the "state of the art" and involved all of the contributors to this volume except Dr. F. Erber, Dr. F. Stewart, and Mr. W. Morehouse who was, however, an active participant in the discussion that followed the presentations. Mr. W. Eilers (Office of Science and Technology, AID) was also a member of the panel. A second panel, recruited by Professor Henry Nau (George Washington University), dealt with the U.N. Conference on Science and Technology for Development. The participants were Ambassador Jean Wilkowski (U.S. Coordinator for the preparations for the conference), E. Brady (Associate Director of the National Bureau of Standards, Department of Commerce), J. Clayman (President, Industrial Union Department, AFL-CIO), Harvey Picker (Dean, School of International Affairs, Columbia University), Herman Pollack (George Washington University), and Donald Robins (Senior Vice President, Singer Sewing Machine Company). The third panel, which focused on the role of science and technology in U.S. foreign policy, was arranged by Ms. Jean Knezo (Congressional Reference Service). The participants included W. Lawrence and F. P. Huddle (Library of Congress). F. Frank (International Relations Committee, U.S. House of Representatives), and Dr. J. Stewart (Consultant on Science and Technology).

The reception accorded to these panels persuaded those of us who were primarily concerned with the academic aspects of this topic that there was a need for a work that would examine the major issues in a systematic manner. Accordingly, we decided to refine and expand our papers so as to produce a volume that would be useful for teaching and research. Recognizing that there were some gaps in our coverage, we agreed that we would invite other scholars to contribute to the volume.

As in any project of this kind, the work quickly assumed a life of its own, one that did not always conform closely to the original design. Nevertheless,

I believe that the final version comes close to fulfilling our original expectations. If it does so, it is due to the efforts of many persons who, in addition to the contributors, helped make this volume possible. Professor Henry Nau made useful suggestions at an early planning stage. My colleagues, Robert Rycroft and Richard Li, in the Technology, Modernization, and International Studies Program at the Graduate School of International Studies, supported my involvement in this project in many ways. Ms. Betsy Brown of Praeger Special Studies proved to be a patient, understanding, and able editor. Mrs. Arline Fink, who acted as executive secretary for the project, deserves my profound thanks for having handled what was sometimes a rather frustrating assignment with her usual good humor and competence. Finally, I wish to thank my graduate research assistant, Ms. Deborah Fox, for her thorough assistance with the index.

Contents

Technology and International Affairs

1

Technology, The Nation State

An Overview

JOSEPH S. SZYLIOWICZ

Introduction

In recent years, technology has emerged as a topic of major concern throughout the world. Debates about whether it is a positive force that will lead to the betterment of the human condition, or whether it is responsible for such contemporary evils as resource depletion, environmental degradation, and the threat of nuclear annihilation, are commonplace. Decision makers, however, regard technology from a very pragmatic perspective. For them, technology is a major factor of national power, a tool that can be used to further a state's domestic and foreign policy objectives whether these be military strength, economic growth, or the elimination of poverty, and to enhance its prestige in the world community.

Such an attitude is relatively new, although technology is as old as civilization and has always profoundly affected man and his societies, local, national, and international. Indeed, without technology one cannot conceive of what the late Jacob Bronowski called the ascent of man for the great historical epochs — the "irrigation revolution," the "industrial revolution," and the "post-industrial age," were, and are, essentially fueled by technological change. In each case, particular technologies profoundly influenced social patterns, values, and economic and political relationships.

Civilization, for example, had its origins in such river valleys as the Nile, which gave rise to the glories of ancient Egypt and the Tigris and Euphrates where Mesopotamian culture arose. Each of these was made possible by technology — the application of knowledge to forecast the annual floods, to measure them, and to harness them through canals and waterworks. The result was increased production, which permitted the development of complex social systems characterized by bureaucracies, armies, social classes, legal codes, and new cultural organizations.

Not until the nineteenth century, however, did an awareness of the centrality of science and technology to national goals become widespread. Scientists had, at least from the time of Sir Frances Bacon, argued for the importance of science, but science and technology gained widespread recognition as essential ingredients of national power only during the nineteenth century as rapid German industrialization became a subject of admiration and envy. World War I, and especially World War II, further accelerated the recognition of the linkage between science, technology, and national and international affairs. Indeed, World War II represents a major threshhold, for that war was fought not only on battlefields, but also in the laboratories and institutes where scientists and engineers struggled to develop code machines, radar, rockets, bomb sights, and, of course, the atomic bomb.[1] Indeed, it may not be an oversimplification to suggest that if the battle of Waterloo was won on the playing field of Eton, the outcome of World War II was decided at Peenemunde, Los Alamos, and Bawdsey.

Today, all governments recognize the centrality of technology and seek to harness it for national purposes. It is widely accepted that whether one is concerned with poverty, overpopulation, productivity, pollution, or any other policy issue, technology represents an important variable. Moreover, science and technology have acquired unprecedented prestige and any state that aspires to eminence in the world community must, therefore, emphasize its commitment to scientific and technological advancement.

Modern technologies, however, contain within themselves not only unprecedented promise for resolving social problems, but also great dangers for mankind. Nuclear power has the potential for plentiful energy or Armageddon, genetic engineering, new standards of health, or the creation of new virulent strains of disease. And, the nature of modern technologies is such that damaging impacts can be diffused more and more quickly over larger and larger areas. Thus, the benefits and costs of new technologies are imposed upon ever larger populations at the same time that the time available to deal with the unanticipated consequences is shrinking and the costs of doing so are increasing. The Aswan High Dam in Egypt and the Green Revolution illustrate this point only too vividly.

Not surprisingly, therefore, a new attitude toward science and technology has emerged in recent years. The widely held view that technological advance was the key to progress and the betterment of the human condition, that technology would bring people closer together, overcome shortages, extend man's control over the environment so that all human difficulties could be alleviated significantly, has been replaced by very different perspectives. One that is quite common regards technology as an evil force, which is responsible for many of the problems confronting developed and developing countries and the world community as a whole. And, some argue that technology is a force that has become autonomous and that is imposing its values and logic on societies everywhere.

Most scholars and practitioners, however, would not subscribe to such a gloomy view. They would not agree that technology is an exogenous force over which man has little or no control. Rather (and this is the position of the contributors to this volume), the consequences of technological innovation are determined by particular social, political, and cultural variables within any particular society. One has only to compare ancient Egypt with ancient China or with Assyria, societies with essentially the same technological base, but with quite dissimilar institutions and values, to recognize the role of social and cultural factors.

Societies have even been known to reject technologies — the Ottomans did not permit printing in Turkish or Arabic until the eighteenth century, even though the printing press had been introduced into the Empire as early as 1493 by Jewish refugees from Spain. An even more dramatic example is that of the Japanese, who deliberately chose to stop using guns in medieval times. The gun arrived in Japan in 1453 and diffused so quickly that by 1575 it had become a decisive factor in warfare. Yet, less than 50 years later, the Japanese had reverted to bows and arrows because the nobility had decided that the democratization of warfare, which the gun made possible, posed severe threats to the stability of the existing social structure. Accordingly, smiths were subsidized *not* to manufacture the weapon; those guns that were produced were purchased by the government and used for ceremonial purposes only[2]. And, in modern times, we have the decision by the United States not to build the SST.

These examples illustrate the point that the impact that technology has upon society is determined primarily by the character of the society itself. Technology is not some sort of external independent force, rather an interactive relationship exists between technology and any social system — just as technology influences the culture and social organization of a society, so does the culture, values, and distribution of political power within a society shape the ways in which technology will impact upon and be utilized by a particular social system. This perspective does not mean, however, that technologies are neutral: on the contrary, they yield both positive and negative results simultaneously and are the carriers of values and of patterns of social organizations

Accordingly, the fundamental issue for societies everywhere is how to develop institutions and how to choose and implement policies that regulate the ways in which science and technology impact upon human affairs. The contemporary concern with the costs and benefits associated with technological change, with the distribution of these costs and benefits, and with the ways in which decisions involving technology are made, is not only legitimate but long overdue, for science and technology are not panaceas. There is no such thing as a "Technological Fix" for social problems (e.g., birth control devices to halt population growth) because the very act of solving one problem inexorably and inevitably creates a range of new problems whose solutions are each as difficult as the original. What remains to be understood and assessed, therefore, is not the promise of science and technology, but rather how that promise can be implemented in a manner that genuinely promotes human welfare.

All of these considerations have led to a proliferation of academic and governmental activities in this field and experts in many disciplines are trying to identify more precisely the impact of technology on human affairs and the ways in which techniques and mechanisms for the control of technology for productive ends can be implemented and improved at the national and international levels. Courses have proliferated, research programs have been established, professional societies have been formed, newsletters and journals circulate, conferences and meetings abound, and scholarly literature grows exponentially. National governments and international agencies have contributed greatly to these activities. They, too, have sponsored numerous research projects, symposia, workshops, and conferences, including the United Nations Conference on Science and Technology for Development (UNCSTD), held in Vienna in the summer of 1979. Numerous other major international conferences dealing with technology in one form or another are scheduled for this decade.

Although these activities have greatly enhanced our understanding of the general relationship between technology and human affairs and have permitted us to identify important issues therein, much of the literature is normative and hortatory, and what empirical work is available is largely limited to specific issue areas. The process of integration has barely begun and no paradigm exists that would permit us to analyze in a theoretically based way the relationship between technology and international affairs.

Given the analytical difficulties involved, it is easy to understand why this should be the case. First, there is no agreement on basic concepts. As Victor Basiuk points out in Chapter 8, international relations theorists disagree on what constitutes the international system. Second, the relationships involved are quite complex and are often affected by feedbacks. Third, we are not dealing with unicausality — science and technology are but one of a number of interrelated factors that have brought about such profound changes in the contemporary world. Fourth, to identify and measure the changes that occur is no simple matter.[3]

Under these circumstances it behooves one to be modest. Accordingly, in the rest of this chapter, it is my aim to demonstrate how the specific analyses that follow are interrelated, and to provide some background information on areas of contemporary concern that are not dealt with in detail later. The framework that I shall use is based on the following perspective. Technological developments occur in particular contexts (primarily within nation-states) and produce international outcomes that involve different units and levels of analysis. These can roughly be divided into two general categories. The first category involves the structure and functioning of the system itself — the number and type of actors, their capabilities, and the distribution of power within the system. The second category involves subsystemic relationships such as the major cleavages and alignments that characterize the contemporary international scene. And, technology can be regarded as a dependent variable, since technological developments are themselves often influenced by such variables as the values of the international system and its particular cleavages.

Thus, I begin by considering basic concepts and definitions, the nature and relationship of science and technology, and how they function as societal forces in the modern world. I then turn to a consideration of the nation-state as a major actor, which has been both affected by scientific and technological developments and is the source of these developments. Then, I move to the international system. Here, I shall consider first the role of technology at the subsystemic level, in terms of the major alignments, and then the role of technology at the global level.

I have organized the contributions that follow so as to complement this approach. The first section analyzes the changes that science and technology have wrought upon the developed countries and the problems and opportunities that these societies confront. The second section considers the state of science and technology in the developing countries and the prospects for the establishment of an effective scientific and technological capability. The final section analyzes the role of technology at the international level.

The Technological Process

The power of the new technology derives from its scientific base. Until recently, science and technology were separate activities, each of which was carried on more or less independently of the other. The distinction between the two may be highlighted by the simple but useful definition of science as "know why" and technology as "know how."

Each is practiced by different kinds of professionals and takes place in different settings (the scientist tends to work in universities, whereas technologists are usually found in the productive sector), operates according to different norms, and produces very different results (the scientist publishes his research findings, the technologist tends to take out patents).

Historically, the relationship between these two human activities has been quite limited, though technology has often been responsible for scientific progress. It was technology that provided science with the tools and equipment required to make many of the major scientific discoveries that have so shaped our lives. Galileo's work, for example, would have been impossible without refinements in the manufacture of telescopes. Ironically, therefore, science cannot exist without technology, for tools are required to test hypotheses, but technology can exist without a scientific base, as was the case in all societies before science became institutionalized in Western Europe in the eighteenth century and as remains the case in many underdeveloped countries today.

Increasingly, however, science and technology have become coupled in a synergistic manner and, as technology has become more and more science based, so the process of invention has changed. Invention no longer takes place in somebody's garage. It is seldom the result of empirical work by a solitary genius. Rather, it is the product of organized and systematic effort. Science and technology have become institutionalized in all the developed countries, and today research and development are activities carried on in well-equipped laboratories and institutes by highly trained professionals who are specialists in particular areas. Research and development is recognized as a valuable activity by both industry and government, and large amounts of resources are allocated to it everywhere. Overall, an estimated 3 million persons are engaged in R & D throughout the world and about $150 billion is devoted to it annually.

Accompanying the changing relationship between science and technology and the resultant shift in the nature of technology has been a redefinition of the concept itself. Until recent times, the definition of technology was quite restricted, the word referred to the "practical arts." Indeed, technology was not regarded as a particularly significant phenomenon, but simply as a synonym for industry and equipment. The changing character and increasing complexity of technology has led to a marked growth in the concept. This trend can be illustrated by simply contrasting the definition given by *Webster's International Dictionary*. In 1909 it was "industrial science, the science of systematic knowledge of industrial art," whereas in 1961 the definition expanded to "the totality of means employed by people to provide themselves with the object of material culture."

Practically every scholar who is concerned with analyzing technology and every practitioner who has attempted to devise policies to promote, control, or harness it accepts a broad definition of this sort because technology involves far more than "hardware" — physical tools and machines. It also encompasses "software" — those bodies of skills and knowledge that are required to ensure the effective and efficient operation of the "machines," as well as to organize particular social activities. Thus, technology has been defined as "any kind of practical know-how" or "any set of standardized repeatable operations that regularly yield predetermined results." Subsumed within such definitions are: (1) machines and tools of all kinds, (2) methods, routines and procedures (e.g., computer programs and algorithms), and (3) patterns of organization and administration.[4]

The difference in these conceptualizations has important implications. Many persons still regard technology as consisting merely of machines. They ignore the extent to which skills and knowledge are an integral part of the process and do not recognize that technology can be person-based as well as process- and product-based. Hence, numerous failures in policies designed to develop, transfer, promote, or diffuse particular technologies have occurred. As one experienced U.S. official has noted in regards to questions of security, trade, investment and foreign aid:[5]

> Failure to appreciate the multiple dimensions of the problems associated with the successful introduction, production, and use of new, innovative applications of technology has been a major reason for the generally ineffectual and frustrating international dialogue on technology policy.

Further complicating national and international efforts to harness technology for specific purposes is the fact that technology is but part of a larger process, which is often referred to as innovation. Innovation can usefully be distinguished from two related concepts, invention and diffusion. Invention may be defined as a new idea that is likely to work. Often, it is developed to the point where it may be patented. Innovation, however, has two separate meanings. In the sociological sense, it applies to the introduction of a technology into an organization for the first time. From the perspective of technology, however, it involves the process whereby an invention is transformed into an item that can be marketed. In this sense, innovation encompasses the various stages of the technological process such as development, design, and testing. Diffusion refers to the spread of an innovation among users. Many practitioners, however, prefer the term "technology transfer" because that phrase implies purpose and direction, that the diffusion of a new product or process involves several critical dimensions including the knowledge required to operate and maintain the new technology. To further complicate the definitional arena, some practitioners make a distinction between horizontal transfer and vertical transfer. The former is a synonym for the innovation or technological process, the latter refers to the transfer of a technology from one country to another or from one industrial sector to another.

Whatever the definition, governments throughout the world are in agreement on the importance of the process whereby ideas get produced and transformed into products and processes. This process is usually referred to as research and development (R & D). Research is frequently divided into two categories: basic and applied. Basic research is defined as those activities that are undertaken in the pursuit of knowledge, applied research as those activities that are goal-oriented. Development involves turning research findings into hardware and software. These activities cannot be separated easily and it is not always possible to categorize research under the rubric of "basic" or "applied." Although R & D is a vital part of the innovation process, other aspects such as engineering design, production, and marketing must also be carried out effectively and efficiently if innovation is to take place.

A number of major research projects have attempted to identify the particularities in this overall process, including the barriers that tend to hinder the transformation of inventions into innovation. Essentially, two basic models have emerged, the "discovery-push" and the "demand-pull." The former emphasizes the importance of R & D, which yields a new invention, which then, in effect, creates a market for itself. This model may be schematically depicted as follows:

Basic Research	Applied Research	Development & Testing	Production	User

Many innovations that have moved along this path have resulted from laboratory discoveries for which no awareness of potential applicatons existed. Nuclear power is one such example.

The second model emphasizes the role of the market, the specific need that exists to be filled. This model, which describes the process by which synthetic rubber was developed, may be depicted as follows:

User Need	Pool of Knowledge	Awareness of Solution	Matching Solution to Need	Critical Evaluation of Technological Solution	Development, Testing, Production

Several projects have shown that this model accounts for about 70 percent of all innovations. However, various studies have also indicated that such innovations tend to be "minor" as opposed to "breakthrough" innovations, most of which resulted from a recognition of a technological opportunity.

FIGURE 1.1   Interrelationships between Horizontal and Vertical Technology Transfers and the National Planning Process

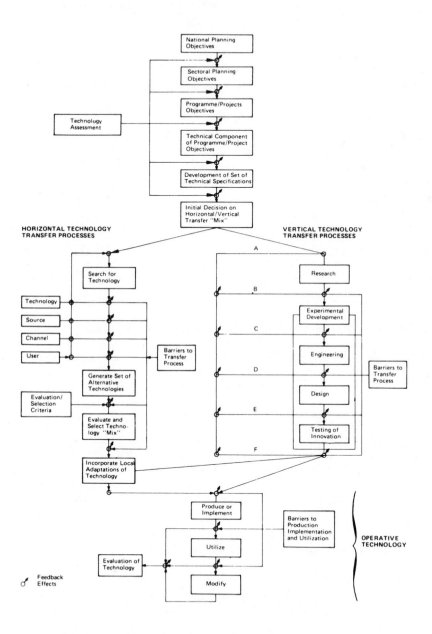

Source: An Introduction to Policy Analysis in Science & Technology, p. 50 (Paris: UNESCO, 1979)

The former category describes those incremental advances that occur daily in industry, the latter to major changes in the "state of the art," such as lasers, integrated circuits, and transistors. One should not, however, assume that the former is not particularly significant, for sometimes the sum total of incremental advances results in a technology that has a greater impact than any "breakthrough" innovation. The development of the jet engine illustrates this point well.[6]

Both of these models, which can be viewed as complementary, suggest the significance of linkages. The technological process involves several stages, each of which is affected by different barriers and facilitators, and each of which must be related to the others. Though often separated in organizations, such functions as research, development, and marketing must be linked for successful innovation to occur. Moreover, to be socially useful and productive the output must be diffused throughout relevant sectors. Here, too, linkages are critical, for many factors, including societal attitudes and values, government policies toward regulation, patents, and taxes can have a positive or negative impact on the innovation process and the diffusion of technology. Hence, many experts stress the importance of linkages between R & D organizations, the government, and the productive sector. The complexities of the process and the interrelationships involved are depicted in Figure 1.1.

Science, Technology and the Nation-State

In terms of the models presented above, one might say that the implicit basis for the formulation of most scientific and technological policies has been the "discovery-push" model. Most attention has been paid to the supply side of technological innovation, to the establishment of R & D organizations and their functioning. Yet, even though developing countries have, in recent years, emphasized the establishment of R & D institutions and pursued a "discovery-push" oriented strategy, these efforts have seldom yielded significant results and have been of minor significance from an international perspective. R & D remains heavily concentrated both in terms of where it is carried out and the sectors that are emphasized. Most research is carried on in the developed countries — the United States, the Soviet Union, West Germany, Japan, France, and Great Britain account for about 85 percent of the $150 billion spent on R & D annually throughout the world; all the developing countries account for less than 3 percent of the total. Overall, the United States accounted, in 1979, for about one-third of the world's total expenditures. Western Europe and Japan also spent about $50 billion annually on R & D. Although the United States accounts for about half of the Western world's total of R & D expenditures, all the Western countries are allocating about the same percentage of their GNP to R & D, between 1.8 percent and

2.3 percent. Only Canada and Italy fall below these figures. No reliable data are available for the Soviet Union and Eastern Europe, but it is generally agreed that they, too, are devoting large amounts of resources to R & D, an estimated 30 percent of the world's total, about $45 billion.[7]

These figures, however, do not reveal the changes that have occurred in recent years. Between 1967 and 1975, United States expenditures did not keep up with inflation and only in the past few years has real growth been resumed. Overall allocations have dropped from 3 percent to 2.3 percent of GNP. The decline in United States R & D financing is at least partly attributable to the rise and equally rapid decline of space R & D. French and British allocations have also dropped, whereas the Japanese and West German expenditures have increased. Soviet and East European bloc expenditures increased by 10 percent a year until the mid-1960s and since then the rate of increase has slowed to about 5 percent per annum.

When viewed sectorally, it is obvious where the priorities are — 24 percent of all R & D expenditures are devoted to military concerns, and the remainder is divided among all other areas, including energy, health, transportation, and agriculture. These figures, however, disguise important intercountry variations. The United States and the Soviet Union invest heavily in military R & D, but only the United Kingdom among the Western countries devotes an equally high percentage, about 50 percent, to national defense. France spends 30 percent, West Germany 14 percent, and Japan 20 percent. Similar variations are to be found in allocations to basic research. The United States, the Soviet Union, and Great Britain devote between 12 and 15 percent to it, primarily because of their emphasis upon military research, which requires large investments in applied research and development. Japan also devotes a low percentage of its expenditures to basic research, whereas France and Germany spend 20 percent or more in this area.

Different institutions carry out research in different countries. In the Soviet Union and Eastern Europe, most basic research takes place in specialized laboratories; in Western countries, it is carried out in universities and most applied research and development, though funded by tax revenues, is carried on in industry. There, the emphasis tends to be on the applied end of the technological process, on the design and development of new products rather than on basic research. And, often the emphasis is upon minor changes rather than major breakthroughs. Increasingly, corporations feel under pressure to minimize risks and to maximize profits, thus further limiting the concern with "breakthrough" innovations. This focus upon the short term rather than the long term has been criticized in many quarters as a major factor that has adversely affected United States technological capabilities. Altogether, private corporations account for between 60 and 70 percent of all R & D in France, Germany, Italy, Japan, Sweden, the United States, and the United Kingdom. R & D, however, is not emphasized to a similar degree in all sectors of industry. In the United States, for example, aerospace, electronics, machinery, computers, automobiles, chemicals, and instrumentation account for about 80 percent of the total.

These patterns have obvious and significant implications from the perspective of the developing countries and of the international system as well. We shall discuss these in detail below. Suffice it here to note that most of the world's R & D is oriented toward military affairs and most of the remainder is relevant to advanced country problems. In industry, the focus is upon technologies that are capital intensive and appropriate for developed economies. Even the R & D that is carried out in such sectors as health and agriculture is also oriented to developed country needs; research is carried out on such diseases as cancer, for example, rather than on sleeping sickness, even though millions more people suffer from the latter than the former throughout the world. Nor are such problems receiving adequate attention in the Third World. A few countries, notably, India, Mexico, and Brazil, are emphasizing R & D, but, here too, resources are often allocated on the basis of the same consideration that influence patterns in the developing countries — power and prestige.

Since R & D is not being carried out by either the developed countries or the developing countries in many critical areas, a new pattern of cooperative research institutes supported by private foundations, international agencies, and foreign governments, is emerging. In agriculture, for example, a new network of 11 institutes based on the well-established International Wheat and Maize Improvement Center in Mexico and the International Rice Research Institute in the Philippines, which played a leading role in the Green Revolution, has been created, with an annual budget of about $100 million. Another network has been launched in the area of tropical diseases and discussions are presently underway concerning a similar arrangement that would specialize in cotton.

As noted earlier, R & D is but part of the technological process and the allocation of resources to this activity at either the national or the international level does not necessarily translate into developmental advance. Indeed, a relatively small number of states in the world today possess the capability to utilize science and technology effectively to achieve particular goals. Such states are the most modern ones, for science and technology are more than tools. In a very real sense, they are the essence of modernity. Development entails, above all, the transformation of societies that cannot deal with change into societies than can do so constructively, societies that possess the ability to make discriminating decisions about technological alternatives and can thus increase the choices available to their citizens.

This kind of perspective underlies the detailed analysis of the relationship between science and the nation-state that J. Schmandt presents in Chapter 2. He carefully distinguishes between the different roles of science (and technology) — as product, as evidence, and as method, and he demonstrates how in each role, science (and/or technology) affects the structure and functioning of the nation-state, how they are responsible for a shift from the "administrative state," a conceptualization advanced by the great German sociologist, Max Weber, to the "scientific state." The former is based on the principle of legal rationality and is characterized by a large, hierarchical, and centralized bureaucracy. The latter, however, is based on scientific rationality and is organized on the basis of decentralized functional networks. In this state:

> . . . government and its agents and institutions (can be seen) less as custodians of stability and more as change agents whose function it is to formulate the implications of technological, economic, and social change in terms that are understandable to the public and to search for appropriate public policy responses to constantly changing conditions. The task is less to administer what is known and more, at least at the frontiers of public service, to manage the unknown. (page 64)

Few states in the contemporary world are close to this ideal type or to that of the "administrative state," for that matter. The majority can be classified as "developing," a general category that includes a very wide variety of states. Hence, some scholars make a distinction between the "Third World" and the "Fourth World," between those countries that have made some progress on the arduous journey to modernization, and those that have barely begun.

According to a recent estimate, about half of the nations of the world fall into the latter category.[8] They remain primarily agricultural and rely on imports to satisfy their limited need for modern technology. Skills and knowledge necessary to maintain and operate these technologies are provided largely by foreigners. Their educational systems are not very developed and there is little emphasis upon science and technology at any level. Some R & D activities are carried on, but with little impact upon the society. Often, there is a brain drain of local experts and professionals of significant proportions. Planning and administrative activities are not well advanced and are carried out according to traditional norms.

Another 40 countries or so fall into the former category. These countries have begun to industrialize by producing consumer and intermediate goods for the domestic market. The costs of this strategy are analyzed by F. Stewart in Chapter 7. Partly because of such critiques, many of these countries are now attempting to combine an import substitution strategy with an emphasis upon exporting some of these goods. All have stressed education and the number of students has increased at all levels, often at the cost of quality. In many cases, however, the educational systems are still unbalanced in terms of scientific and technological manpower needs and of the attitudes and values that are disseminated. Moreover, these countries have attempted to develop a scientific technological capability and to direct research and development efforts towards specific local needs. As a result, an R & D infrastructure exists, but it tends to be characterized by quantitative and qualitative difficulties of various kinds. Human and physical resources are limited, problems of quality are more or less severe, linkages are weak, and most research, especially in the universities, is oriented toward the theoretical rather than the applied, to problems of interest to the international scientific and technologial communities rather than of local relevance. Planning for science and technology is seldom carried out in an integrated and sophisticated manner and little attention is usually paid to screening technologies, to assessing the advantages and disadvantages of alternatives. Most innovation is introduced by foreign companies, local industry invests little in R & D and imports whatever technology is required.

The industrialized countries of both East and West represent a third group. Here we come to the "administrative state." These societies are highly urbanized with a developing service sector and universal primary education. They are sensitive to the importance of science and technology and various bodies have been created to coordinate and promote research and development, to integrate scientific and technological activities into national planning. The emphasis, however, tends to be upon formal administration and legal rationality, and problems in such areas as integration, coordination and implementation are common. About 25 countries belong in this group.

Finally come the societies that are on the threshhold of becoming "scientific states"; there are perhaps a dozen of these. Here, the emphasis is upon the quality of life rather than on economic growth, upon the application of scientific knowledge to resolve problems. The scientific and technological communities play active roles in decision making at various levels and questions involving the emergence of "technocracy," the role of scientists and technologists in the political arena, the ways in which decisions about major technological policy and projects should be made, and how the general public should participate, are issues of general concern and debate. In these countries, research and development are highly developed activities and the bulk of the world's R & D is carried on there.

Since the "scientific state" represents an ideal type, it should not be surprising that even in the most advanced countries, as Schmandt himself notes, scientific and technological policy making has been only partially successful. This topic is expressly addressed in Chapter 3 by Leonard and Gilpin, who analyze scientific and technological activities in the Western countries and find marked differences in policies followed and results obtained. Essentially, the advanced countries have pursued three general strategies. The United States and the Soviet Union have adopted a "broad front" approach, seeking to promote the advance of innovation in all high technology areas. Other countries, such as Sweden, the Netherlands, and Switzerland have traditionally specialized in specific sectors and sought to focus their resources in a particular area of commercial importance. Japan is the exemplar of the third approach, one that is based on the adaptation of technology developed elsewhere.

Although each strategy has its own advantages and disadvantages, it is noteworthy that Japan has successfully promoted technological innovation and has achieved high rates of technological growth, whereas the United Kingdom, which originally attempted to pursue a broad front strategy, has failed to do so despite the presence of a sophisticated scientific base. Leonard and Gilpin suggest that several important lessons can be derived from these cases. In particular, Japan's strong performance can be attributed to such factors as the integration of technological activities into economic planning, the explicit design and implementation of policies to promote growth, and an emphasis upon the demand rather than the supply side of the technological process, upon the links between users and producers. The British experience further underscores the significance of these considerations. There, linkages between the government, industry, and the educational system were very weak and the government emphasized R & D in high technology areas and sponsored commercial development of particular technologies, often doing so prematurely.

Such findings are particularly significant given the situation in the United States where there is widespread recognition that the era of the 1980s poses new social and economic problems — growth will be more difficult to achieve and competition is likely to be more severe than ever before. Although foreign trade has represented a relatively minor aspect of its overall economic activity, the United States position has deteriorated steadily and significantly in recent years. Its balance of trade has shifted dramatically in an adverse direction in the past decade, not only because of higher energy and raw material costs, but primarily because of the declining competitiveness of major industrial sectors. Steel, textiles, consumer electronics, and automobiles, among others, have all proven unable to compete with foreign manufacturers, owing primarily to falling productivity, which, in turn, is a function of numerous factors, including inadequate investments in technology and developments in other countries and in their relationship with the United States, which have profoundly affected the United States economy. We shall explore these issues below. Suffice it here to note the significance of such relationships, that the competition from other developed countries and some advanced developing countries have forced every modern state to attempt to develop an appropriate policy response in the industrial arena.

These developments are probably inevitable to some degree — the emergence of other developed countries as major producers of new technologies may simply be the result of their successful recovery from the wreckage of World War II. As one scholar recently noted:

> . . . the United States has little cause for alarm from a reduced technological gap so long as the reduction in the gap results from an increase in other nation's abilities to innovate *rather than a decrease in our own innovative capabilities.*[9] (Author's emphasis.)

That, of course, is the critical issue, whether the United States capacity to innovate has, in fact, deteriorated. After all, as the Japanese case so vividly demonstrates, the ability to produce technology and the ability to harness technology to achieve national goals are not necessarily linked. If the pool of international technology is growing as a result of increased technological productivity by other countries, then the situation is not necessarily detrimental to United States interests because costs and risk are being assumed abroad.

Nevertheless, there is widespread agreement that existing patterns of technological innovation in the United States are not satisfactory and the question of how to achieve revitalization has emerged as a topic of major contemporary concern. Numerous recommendations for policy changes in such areas as taxation, regulation, patent law, antitrust law, and support for R & D have been advanced.[10] That so many areas are involved reflects the complexities of the technological process and the difficulties in designing coherent and integrated policies.

Such policies can be developed only if a specific strategy is selected. Essentially, three general orientations are available to the United States. It could become a *rentier,* living off foreign investments, and is, in fact, behaving increasingly in this fashion. Or, it could become more isolationist, more protectionist, supporting industries that are essentially noncompetitive. Powerful domestic elements favor this strategy, which is welfare- rather than growth-oriented. Or, it could adopt a positive strategy that is designed to rejuvenate the economy. Such a strategy would encounter severe constraints, not least of which is the increasing competition among the developed countries for markets. Nevertheless, this strategy remains the most desirable. Implementing it, however, entails difficult and complex choices in terms of the allocation of resources to science as contrasted to technology, to competing sectors, and to different aspects of the technological process itself. Much, however, can be learned from the achievements and failures of other countries and adapted to the United States context, and it is possible to design policies that will lead to the development of a flexible economy, one in which productive factors shift quickly and easily into the most dynamic sectors. The commitment to and successful implementation of such a strategy is not only desirable, but essential, if the aspirations of the American people and, indeed, of peoples throughout the world, are to be fulfilled in years to come.

Such considerations reflect the extent to which scientific and technological policy making within nation-states are linked to external considerations at both the subsystemic and the global level. Hence, it is often difficult for a state to harmonize its domestic interests with its foreign policy objectives. Leonard and Gilpin summarize this point as follows:

> . . . nearly every action taken by the state to encourage preservation, adaptation, or innovation in the industrial sector is bound to be seen by some foreign groups as tantamount to the provision of an artificial advantage in the international market.

> Thus, under the rubric of a state's industrial policy, there is liable to be a constant tension between the state's welfare goals, its international competitiveness goals, and its international relations goals: it is unlikely today that one policy would work towards maximizing all of these goals.

How the United States will resolve this dilemma remains to be seen, but it is ironic that it should be regarded as a country in which technological innovation is lagging because in the post-World War II era, the United States was widely regarded as the one country that had developed and implemented a successful strategy for harnessing science and technology and, in the 1960s, the Europeans were expressing widespread concern about the "technology gap" between themselves and the United States.

As C. Wright points out in Chapter 4, however, that simply was not a realistic fear. His analysis of United States science and technology policy reveals an absence of any overall strategy. By and large, United States technology policy was the by-product of policies in specific areas such as national defense. Decisions were, and are, made in a fragmented and, often, ad hoc fashion by numerous agencies, each with its own mission and clients. Little attention is paid to rationalizing the process, to the design and implementation of a coherent and integrated technology policy that is consonant with, and supported by, policies in related areas. Moreover, little attention has been paid to the requirements that the changing international environment poses. Thus, questions of domestic competition, for example, play a far greater role in shaping the structure of United States industry than the functional requirements for effective competition in the international arena. Similarly, the orientation of policies toward the scientific and technological system has been toward the maintenance of existing patterns rather than to assuring its capacity to function effectively in an international context. The need to move from such diffused, incremental policy making toward more rational, comprehensive planning should be obvious if the United States is to become the kind of flexible and creative society that can thrive in an increasingly competitive environment.

Such concerns are not reserved for the developed countries. Questions of innovation, of the appropriate policies and mechanisms for the production and diffusion of technology, are at least equally worrisome to the developing countries — the first two categories of the typology presented above. Our discussion there demonstrated that science and technology contribute little to the resolution of problems in those societies and, in fact, may aggravate them. Despite efforts of varying intensity, these countries possess a very limited R & D capacity, with serious qualitative problems that are more or less common. Although some R & D activities may be carried out, they often do not constitute an organic part of an integrated and productive technological system. As a result, numerous inventions lie on the shelves of R & D organizations, inventions that have never successfully penetrated the marketplace and contributed in any way to the achievement of the country's goals.

In Chapter 5, Professor J. Montgomery focuses particularly on the role that science and technology can play in improving the quality of people's lives. He sensitizes us to the limitations of policies in this field and to the physical and human constraints upon their development and implementation that exist in most developing countries. His analysis demonstrates that diffusion represents the most signifiant aspect of the technological process for development; hence he stresses questions of linkages and discusses various innovative arrangements, all of which are designed to optimize diffusion by creating a partnership between government and entrepreneurs. His presentation highlights the importance of the demand side of the technological process.

The results of a recent major international study of science and technology policy strongly support this conclusion.[11] It found that policy instruments that strengthened the demand side were of greater utility in stimulating indigenous scientific and technological activities than those that were aimed at the supply side. Hence, any government wishing to promote the technological process would be well advised to consider the potential contribution of such instruments as the establishment of industrial priorities and programs, the pattern of industrial financing, government purchasing policies and the like.

In his rich analysis, Montgomery also explicitly raises an important issue which, enough often neglected in discussions of technology policies, should be accorded a central role: the values that should be promoted and the ethical concerns involved. Technology, as noted earlier, promotes some values, and undercuts others; the very decision to harness science and technology for development presents a profound dilemma which Montgomery phrases as follows:

> . . . the introduction of science and technology either immediately benefits certain individuals or groups whose values are most compatible with the behavior

required in 'modern society' or else it demands changes in the value structure of individuals and groups that desire to take advantage of the new opportunities for self improvement. Is science and technology, therefore, to serve the privileged few whose values and resources are compatible? Or, on the other hand, is it to manipulate divergent values so that others may take advantage of the opportunities it creates?

How to deal with such a situation remains a complex intellectual and empirical task for scholars can contribute little in the way of theory or methodology to assist policy makers who, in any case, usually find themselves in environments that seldom facilitate the process of goal definition and value clarification.

Yet, the issue must be confronted. Accordingly, Montgomery suggests that scientists should assume new roles, that they should become more activist and work to inject value considerations into the policy process. He argues that scientists in developing countries, particularly if supported by the international scientific community, can have considerable influence in incorporating such values as equity into many public policy arenas because political leaders customarily delegate important areas of responsibility to them.

While this suggestion possesses considerable merit, as evidenced by its success in at least one instance, one should be careful not to overestimate the contribution that indigenous experts can make in this regard. There are two reasons for this reservation. The first reflects the heterogeneity that characterizes scientists and technologists. Their ranks include a broad spectrum of ideological and value perspectives and interests and seldom are they able to unite on specific policies. Moreover, even when the issue involves the allocation of resources to science and technology, one encounters a variety of views.[12] The second involves the nature of the decision-making process for the role that scientists play and can play will be profoundly influenced by the character of the particular political system in which they function. The use of exile and other repressive measures to keep scientists in line is not unknown in the modern world.

F. Erber deals with this issue in Chapter 6. His analysis of the reasons why the technological process possesses the deficiencies identified above leads to the conclusion that social, political, and economic structures are important determinants of the R & D patterns and of the kinds of interactions that characterize relationships between a government, the productive sectors, and scientific and technological organizations. Above all, he emphasizes the leading role of the political, pointing out that such goals as maintaining or enhancing the power of particular elites can deeply influence the content and orientation of technology policy. For this reason, existing patterns of technology transfer and relationships with external suppliers represent a power factor which is often a major obstacle to the development and implementation of policies that can bring about the profound changes in the character and orientation of the productive sector that are required to generate demand for indigenous R & D and to develop positive linkages with domestic technology suppliers.

Hence, questions of demand and linkages must be viewed from a broad perspective and attempts to deal with such issues in the absence of a supportive political environment are unlikely to be effective. Moreover, even though some advanced developing countries have been able to implement policies designed to promote domestic R & D, the pressures of existing economic patterns often act as a powerful force limiting the development of an overall technological capability. Erber summarizes this point as follows:

> . . . one should be wary of wishful thinking: science and technology policies are by necessity subordinate to more general policies and to the economic and political conditions which determine the latter and which set the specific priorities to scientific and technological development . . . our analysis suggests that the pattern of economic development in the peripheral countries and the political structures which express such development tend to reduce the priority of a local development of science and technology in the peripheral countries and to restrict the range of product and process locally developed.

Despite their differing perspectives, Montgomery and Erber agree on the importance of establishing an effective and indigenous scientific and technological capability and on the difficulties involved in doing so, although they would disagree on the relative significance of particular variables. Without such a capability, no state can be independent in the contemporary world, no state can implement policies that will permit it to resolve its particular development dilemmas. Indeed, it represents, as was suggested above, the essence of modernity.

The critical character of such a capability, that it represents an essential prerequisite for modernization, can be demonstrated through a consideration of the different kinds of technology transfer that can occur.[13] A country may choose to engage only in the simplest form of transfer, "pseudo" or "imitative" transfer, that is, buying a modern technology and placing it in a new setting with limited attempts to relate it to local conditions. Even with such a strategy, however, numerous choices are involved and its — albeit limited — assimilation and integration is profoundly affected by the absorptive capacity of a society, a variable that has, at its core, the size, scope, and quality of the scientific and technological infrastructure. Moreover, if the technology is not to become obsolescent, R & D inputs will be required because at least some competitors will be engaged in ongoing innovations of products and processes.

Since R & D activities which are overwhelmingly concentrated in advanced societies are not geared to the needs or the economic and social structures of the developing countries, much of the available technology is either wholly or partially inapplicable to their needs. Hence, in recent years, much attention has been paid to the development of "appropriate technology" and most countries are now seeking to operate at a second, more sophisticated level, that of "adaptive transfer," taking an existing technology and modifying it to suit local conditions, needs, and resources. Clearly, for such activities a more developed infrastructure is required.

The third and most advanced level is called "creative transfer." Here, a country identifies a need and then proceeds to develop a specific technology utilizing existing and new hardware and software in novel configurations. Although many of the advanced developing countries have established a capacity to carry out some adaptations, very few have yet been able to develop the kind of innovative capacity required for such activities. (See Erber, Chapter 6.)

Even though a consensus exists on the importance of developing such a capability, no agreement exists on the particulars apart from a growing concern with demand and linkages, as was emphasized above. There is no agreement on the optimal pattern of allocations, nor on the areas that should be emphasized. Specifically, to what degree should science, as contrasted to technology be promoted? Experts disagree quite markedly on this question. Some argue that scientific activities must be promoted, others that it is a luxury. Basically, the debate centers around the need for basic research. There is unanimity about the importance of applied research, but fundamental research is sometimes regarded as a luxury which should be abandoned or at least deemphasized in favor of applied and developmental work. Certainly, the Japanese case demonstrates the feasibility of a technologically oriented strategy. Conversely, the case of England illustrates how even a high level of science does not necessarily lead to technological innovation. The majority of experts, however, believe that every country should engage in some basic research; this position is supported by such international agencies as the U.N., U.N.E.S.C.O., and O.E.C.D.

There are several important reasons for this position, including: (1) the difficulty of distinguishing "pure" from "applied" research and of deciding on "relevance," (2) the danger of losing one's best scientists if restrictive policies are adopted, and (3) the ever-closer coupling of science and technology necessitates scientific knowledge in order to make productive decisions. Nor can one overlook the fact that, today, science has become a prime determinant of a nation's status.

Many of those who support such research do not, however, believe that it should be absolutely autonomous. It should be emphasized only to the extent that it sustains an appropriate educational effort and should be concentrated in sectors such as agriculture and health, which are usually characterized by specific problems about which little or no knowledge is available elsewhere in the world. Such research can stimulate other aspects of the technological process that impact directly upon agriculture and health care, two areas that can contribute most directly to an improved quality of life for millions.[14]

The development of any kind of scientific or technology capability is, as I have emphasized, a major undertaking. From the earlier discussion, it is apparent that the functioning of the technological process is profoundly affected by the character of science and technology, the associated organizational requirements, and the fit between these and the environment. Particularly relevant variables include: (1) the degree of self-containment of the activities (the extent to which outputs have to be utilized by other organizations if survival and growth are to occur), (2) their technological content, (3) their scope (the number of functional domains that have to be spanned to achieve goals), and (4) the extent to which such activities are reversible.

It is generally accepted that the higher the degree of self-containment and reversibility and the lower the level of technological sophistication and of scope, the easier it is for new roles and activities to become institutionalized within a particular social system. However, in the case of science and technology, and especially as far as the latter is concerned, we are dealing with activities that possess precisely the opposite characteristics. "Pure" scientific work is highly self-contained, and is of narrow scope, for the output — scientific knowledge — is disseminated through international journals, and the norms that govern the activity are essentially international ones. Hence, it is conceivable that a highly sophisticated enclave could be institutionalized in a developing society, provided that other requirements such as appropriate leadership and resources are met. As one moves towards the applied end of the scientific continuum, however, and particularly in the case of technology, one finds activities that are not self-contained and that are of very wide scope, for they overlap with such sectors as higher education and the economic system and involve the adoption, diffusion, and assimilation of new concepts, techniques, rules, and values throughout the society.

Any policy for science and technology, therefore, involves considerations of financial and human resources and of planning and management in and among a variety of institutions ranging from government bureaucracies to research institutions to universities and industrial and agricultural organizations. It involves an understanding of the distinction between science and technology, of the unique features of each, of the different social and cultural requirements for their development. In many cases, policy makers have not recognized these distinctions and have, as a result, sponsored or implemented policies that have not only failed to achieve desired goals, but have yielded socially undesirable consequences. Moreover, the institutionalization of such activities involves highly irreversible changes, for if innovation is to become widespread, profound modifications in values and in the basic regulative principles of the social order are usually required. This is not to suggest that traditional values are inherently hostile to scientific and technological activities, but rather to emphasize the point that the institutionalization of science and technology requires the widespread acceptance of such attitudes and values as an emphasis on analytical thought and scientific rationality, a willingness to experiment and to modify existing practices and institutions to accommodate new findings, and a willingness to extend the application of scientific thought to all aspects of human life so as to deal with the impact of technological change in a constructive manner. Indeed, one expert has argued that the health and level of any country's scientific capabilities is ultimately determined by the degree of "public knowledge," the scientific sophistication of the populace as a whole.

In many developing countries, however, there is no public awareness of the importance of science — science remains a foreign import. This is an extremely significant point. Science and technology, as emphasized earlier, are, above all, social and cultural activities that are profoundly affected by the context in which they operate and the sad state of contemporary scientific and technological activities, the inadequate numbers and quality of the scientists and technologists in many countries is ultimately due to the fact that science has not been internalized by the society; it has never become an area of critical concern.

Accordingly, the present condition and future prospects of science and technology throughout the Third World will be determined not only by the pattern of existing activities, but also by the ways in which the scientific and technological infrastructure is linked to other major institutional orders, and, most importantly, to the legitimacy accorded to these activities as well. The variety of new roles and role sets that are a part of the technological process must become accepted as valued activities that contribute directly to the well being of the society. In other words, the environment must be supportive and encouraging to the requisite activities and to their practitioners. These roles must gain legitimacy in the eyes of the critical elites and perhaps even the population at large, if they are to endure and flourish.

Under these circumstances, a considerable amount of political power will almost inevitably have to be expended to foster the growth and development of the technological process in what are often hostile environments; not all societies possess either the orientation or the resources required for such an effort. The generation of the kind of orientation required obviously is no easy matter and will involve difficult and critical decisions. Yet, unless such policies can be adopted, it is not likely that the promise that science and technology hold can be achieved. The promise will be just that, a promise. In the final analysis, we are dealing with basic questions of politics and ultimately with the wisdom and efficacy of political leadership.

We are also dealing with fundamental ethical and value questions, for it is obvious that if scientific and technological activities are to be institutionalized within a society, profound cultural and value changes may well be required. Some persons argue against this orientation. They suggest that it is Western or "Occidental" science and technology that possesses the disruptive characteristics, that what is required is the development of a culturally based science and technology that is consonant with traditional values.[15]

It is difficult to conceive of such a development, particularly if one accepts the analysis of J. Schmandt in Chapter 2 of the different roles that science (and technology) play in society — as product, evidence, and method. In any event, this debate returns value and ethical questions to the forefront where they belong. Professor Montgomery, who in Chapter 5 addresses himself explicitly to such questions, captures the essence of the dilemma confronting decision makers throughout the world when he writes:

> The greater a nation's dependence on technology for achieving its collective goals . . . the more serious is the potential ethical challenge of scientific activity.

How different societies will respond to such a challenge is, of course, unclear. What is clear, however, is that the development of an indigenous scientific and technological infrastructure and its orientations to serving the needs of the citizenry of a particular state is intimately connected with policies concerning the importation of technology and the role of powerful external actors such as multinational corporations and the like. In other words, scientific and technological activities within a state are, as noted above, linked to and affected by the character of the international system.

Technology and the International System

While discussing scientific and technological activities in developed and developing countries, I alluded to the ways in which such external variables as the degree of international competitiveness and other characteristics of the global economy, including the types and costs of technologies that are available, impact upon nation-states and the ways in which decisions made within nation-states have implications for their relationships with other states and upon other dimensions of the international system as well. In Chapter 8, Basiuk seeks to analyze these relationships systematically, but his effort to do so is greatly complicated by the fact that there is no agreement among theorists of international relations as to just what elements comprise the international system. Accordingly, he opts for an eclectic approach and considers the ways in which technology has affected such dimensions as the number and type of actors, relations among them, basic regulatory principles, and, especially, the impact of technological developments upon the distribution of power within the system. He notes that technological innovations have always heavily influenced the power of individual states and the structure of the international system. Advances in nuclear weaponry illustrate this process vividly, for they led to the emergence of the superpowers and bipolarity. These developments, however, have caused a basic change in the role of power in the modern world, for the costs of using military force have increased. Hence, the importance of military power has been reduced whereas the political, economic, and technological aspects of national power have been enhanced. Moreover, the overall significance of power has diminished so that the contemporary international system remains bipolar only in terms of military power; in terms of all other power factors, Basiuk concludes, we are living in a multipolar world.

Because the United States and the Soviet Union remain the dominant military powers, the pattern of relations between East and West, between the United States and its allies on the one hand, and the Soviet Union and Eastern Europe on the other, is critical. In this relationship, technology plays an important role, as we shall see below. It also does so in other major alignments within the contemporary international system. Those are, first, relations between the developed and the developing countries — between, to use the jargonistic but convenient phrase, North and South; second, interactions among the developed countries; third, an emerging relationship that is garnering considerable attention because of its potential, that between the developing countries themselves. In the pages that follow I shall discuss the role of technology in each of these sets of subsystemic relationships in turn and then analyze the role and impact of technology at the global level itself.

Considerations of national security profoundly influence the role of technology in East-West relations. Western policy has been based on two principles, first, the strategic doctrine of "mutually assured destruction," second, that cultural, economic, and scientific exchanges be promoted so as to increase the scope and scale of the contacts between East and West. Through such linkages, mutual suspicions would hopefully be reduced and the security climate improved.

Trade involving technology has become quite extensive in recent years because the Eastern bloc countries have not succeeded in achieving any of their economic goals and, indeed, actively seek access to Western technologies in order to improve their economic performance in numerous fields. In fact, it may be no exaggeration to suggest that the Soviet and East European economies have reached a point where significant amounts of Western technology and financial resources are required if they are to meet the ever-increasing demands placed upon them by their citizens.

Numerous Western firms, often supported by their governments, have eagerly attempted to penetrate this large and potentially lucrative market and a large number of arrangements for the transfer of technology to East European countries and the Soviet Union have been completed. At present, for example, many large facilities are either under construction or in the planning stage, including steel mills, aluminum plants, chemical plants, and a huge natural gas pipeline that will link Siberia and Western Europe.

Many projects have been financed by Western banks and governments and every East European country has incurred a large debt to the West. This situation poses several major issues that deserve careful consideration. In a narrow economic sense, it is highly likely, for example, that these debts can be reduced only if the East European countries succeed in increasing exports at the expense of Western industries. More basic, however, is the relationship between the technology and capital flows and the issue of Western security. Such activities cannot be considered apart from the foreign and security policy objectives of both East and West. What is, for example, the contribution to the Soviet Union's military capability that results from the export of modern chemical or communications or automotive technology? What levels and kinds of technology should be transferred to Eastern Europe and/or the Soviet Union? What are the tradeoffs between Western economic benefits and Western security in each case? Clearly the United States has attempted to deal with such complex questions through a system of export controls designed to protect its technological lead in military relevant areas; it has sought to limit the sales of sophisticated electronic and other equipment with military applications to the Eastern bloc. Moreover, the United States has attempted to link its technology export policy to the Soviet Union's foreign policy behavior. The Soviet invasion of Afghanistan in December 1979 provoked Washington to curb its exports of high technology and in the course of the crisis over Poland in December 1980, Washington warned Moscow that an invasion of Poland would cause it to incur severe costs in various areas including that of high technology flows from the West.[16]

Attempts to balance the relative advantages and disadvantages of different patterns of technology flows between East and West are extremely difficult, especially since one must weigh the potential economic benefits that are likely to accrue to the exporting nations as well as the overall security gains that may result from less hostile East-West relations. Such calculations are further complicated by the need to analyze the technological and other developments that may occur in the West, as well as present and potential Soviet goals, motivations, and intentions. Moreover, one must assess the potential for differential uses of a technology. Though transferred for a specific purpose, a technology may be used in another manner so as to enhance the East's military capabilities. A second difficulty involves estimating the "seepage" of the technology, the ability of the receiver to modify it, to improve it, and thus to narrow the technological gap.[17]

Questions of East-West technology flows are inextricably linked to United States relations with its allies because they carry on a large part of the total East-West trade in technology. This further complicates the issue, for policies by the different Western states in this area have been marked more by competition than cooperation. As one analyst noted recently:[18]

> To date, rather than posing searching questions about East-West trade, the democracies have encouraged individual businesses to compete for trade deals that may profoundly threaten the security of the West . . . It is surely time to establish guidelines, before the democracies, out of greed and lack of will, further enrich their adversaries.

In this, as in so many other areas involving the developed countries, delicate issues arise, for the advanced countries are essentially allies who trade heavily with each other and at the same time compete with each other in the same sectors. Moreover, each developed nation seeks to use science and technology to enhance its competitive position and has pursued different strategies and developed different mechanisms to achieve this end. And, as we have noted above, the result, inevitably, is tension between domestic and foreign policy goals.

Questions of technology permeate almost every aspect of relations among the developed countries and, ironically, either the development and implementation of successful strategies that accelerate the rate of industrial innovation or the failure to do so can have adverse consequences for the other developed countries of the Western world. The United States, as noted earlier, may opt for a *rentier* or for a more isolationist stance to deal with its deteriorating technological position. Such a policy would have drastic consequences for Western Europe and Japan, and would profoundly alter the nature of United States economic and political relationships with the rest of the world. Conversely, Japan's success is widely regarded as having occurred at the expense of her allies and trading partners. And, the rush to develop and implement policies for industrial innovation by individual democracies is viewed with increasing suspicion by its allies and trading partners. (See Leonard and Gilpin, Chapter 3.)

How to deal with the complex of political and economic issues involved in a way that enhances cooperation and promotes stability, how to make intelligent tradeoffs between internal and external policy considerations, between the short and long term, are problems that will tax the abilities of the most sophisticated decision makers.

Moreoever, the issue of economic rivalry among the advanced countries can easily be aggravated by the increasing salience of the newly industrializing countries — Hong Kong, Mexico, Taiwan, Singapore, and Brazil — within the international economic system. These six are now successfully penetrating developing country markets and the challenge, a development that is already troubling Western policy makers, is likely to grow as other countries industrialize. The problem of how to accommodate the growing stream of finished products from the South has already become evident in such areas as textiles, computer chips, and steel, and may soon expand to include petrochemicals, aluminum, and other sectors. At present, the developing countries' share of global industrial production amounts to about 8 percent. The stated goal, by the year 2000, is 25 percent. Even if the developing countries' share doubles instead of triples, it is not clear how the international economy will accommodate the new exports, or how particular states will react. Certainly, the issue raises many difficult and profound questions concerning the future direction of industrial and technology policy in and among the developed countries, questions that once again require a high degree of cooperation, not only among the developed countries but between them and the "South" as well.

Already, questions involving North and South occupy a prominent place on the international agenda. Here, however, the fundamental issue is not one of economic rivalry, of protecting markets and industries. Technology flows are very different, they are essentially one way, from North to South. The crux of the relationship here is not technological competition but rather technology transfer. Thus, a whole new set of issues arises such as the "appropriateness" of the available technologies, the power of external actors, such as multinational corporations in transferring technology, the availability of technology on reasonable terms, and the proper policy towards imported technology.

The debate over these issues is based either implicitly or explicitly on three dominant paradigms. The traditional one emphasizes an orthodox view of economic development. It considers the continuing integration of the developing countries into the existing international system as desirable. It tends to emphasize the positive functions of multinational corporations (MNC,) as agents of technology transfer who can mobilize, develop, and apply new technologies in a creative and catalytic manner throughout the Third World. The importance of proprietary rights to technology is recognized and the existing international system is not regarded as inherently exploitative. The need to develop scientific and technological institutions in the developing countries is accepted. The focus, however, is upon traditional approaches to economic development, upon accelerating GNP growth through greater emphasis on exports, and through the elimination of those factors that distort genuine market forces in such areas as capital and labor costs.

In the decades following World War II, many developing countries implemented strategies based on this paradigm, and in Chapter 7 Professor F. Stewart analyzes the results. Overall, the strategy was quite successful. A modern industrial sector did develop that was oriented towards reproducing imports; hence, she rightly notes that the common label for the strategy, import substitution, is something of a misnomer. Moreover, the laissez faire attitude towards technology coupled with tax and tariff incentives of various kinds led to a large inflow of technology into these countries.

The costs, however, proved to be very high. The hopes that a growing industrial sector could generate sufficient employment opportunities for rapidly increasing populations were ephemeral, a dual economy emerged with marked and growing disparities between the small modern sector and the large, traditional, primarily agricultural one, which was largely neglected, and imported technologies proved to be inappropriate as well as extremely expensive.

As a result of this experience, a new paradigm emerged that argued that development can be achieved only if the present international order is transformed into one more favorably disposed towards the needs of developing countries. Technology plays a central role in this scheme because it is regarded as the major instrument through which neo-colonialist patterns of dominance by the center over the periphery are maintained. Technology has replaced naked military power though, with equally destructive implications for developing countries. Because modern technology is essentially developed in and controlled by the North, its nature and the ways in which it is transferred have created a situation which benefits the developed countries and prevents the achievements of genuine development. Being characterized by capital intensity and infrastructural requirements which cannot be met without creating major new problems, it is inappropriate, given resource endowments, and it also exacts such high social and cultural costs as aggravating unemployment, overurbanization, social inequality, and other social pathologies, including the destruction of traditional cultures in the process. Nor should one overlook the high financial costs involved in acquiring these technologies, costs that represent a major drain upon fragile economies and increase inexorably because of the power differentials between buyers and sellers. Moreover, indigenous scientific and technological capabilities cannot be developed in this context because the domestic economy organizes itself to supply local elites with the same goods that are available in developed countries. Thus, there is no demand for local R & D, no incentives for developing local technologies, and whatever science and technology does exist within the society is "marginalized," and many scientists and technologists choose, therefore, to join the brain drain. The international system perpetuates this situation, which cannot be corrected unless the linkages that bind the countries of the periphery to the center are eliminated. In short, existing technology flows between North and South strengthen and perpetuate the pattern of inequality and Northern domination over the South. A new colonial system has been created, which, given the growing salience of technological innovation for power and wealth, is leading to ever-greater inequalities between the center and the countries of the periphery. Erber's analysis in Chapter 6 is essentially derived from this paradigm, and he therefore concludes:

> . . . the pattern of economic development of the peripheral countries and the political structures which express such development tend to reduce the priority of a local development of science and technology . . . and to restrict the range

of products and processes locally developed . . . (the) range of action seems to be structurally limited, unless considerable changes in the international system and in the peripheral countries pattern of development supervene . . .

Recently, a third paradigm has emerged that recognizes the significance of existing patterns of global R & D and the power of the multinational corporations, but which emphasizes the critical role of internal forces and the factors in determining the ability of a society to utilize science and technology productively. Technological backwardness is not a condition that has been imposed on the developing countries by hostile external factors; it is a condition that has evolved over time but that need not be allowed to continue. But, massive infusions of technology or other resources are not sufficient to reverse that condition. What is required is the establishment and development of a scientific and technological infrastructure that permits the society to negotiate wisely for specific foreign technologies, to adapt them to meet local needs where necessary and to develop indigenous technologies where none are available externally. To achieve this goal, considerable external support will be required and a new relationship between North and South that facilitates the achievement of new patterns of development, patterns that emphasize such dimensions as equity and the meeting of basic human needs, is desirable. The parimary focus, however, must be upon the domestic scene, upon the profound internal changes that are essential for the creation of an effective scientific and technological capability.

Adherents of each of the latter two paradigms advocate very different strategies toward the importation of foreign technology. Those who regard technology flows as a means of perpetuating dependence tend to argue strongly for a delinking strategy. Those who focus on internal factors advocate continuing reliance on foreign technologies, but in a manner that minimizes their costs, social as well as economic. In Chapter 7 F. Stewart points out that each of these approaches possesses advantages and disadvantages. In her words:

> The first strategy requires a much more radical break with the past — and may involve substantial costs in the short to medium term. A strategy which continues to rely primarily on foreign technology as the basis for local innovations inevitably leads to a continued locking in to the patterns of development of the advanced countries. With some local adaptations, both production and consumption patterns are likely to follow developments in the industrial countries. A more radical delinking of technology is required for more autonomous development.

How any particular country assesses the costs and benefits of "autonomous development" depends on numerous variables, including the quality of the indigenous scientific and technological infrastructure, the distribution of power within the society, and the orientation of the political elite.

Professor Stewart also notes that the very concept of North-South may well be meaningless because important elements within the developing countries benefit from the status quo and are allied with those in the North who also oppose change in existing patterns of technology flows. In order to deal with this problem, she suggests that attention be focused on just who enjoys the benefits and who pays the costs of various technological activities and that, since the underprivileged sectors benefit but little from present patterns, a new strategy of development is required that will promote the welfare of the large and often powerless group.

Because of such concerns, many are now advocating a strategy oriented to meeting basic human needs. This approach, however, like the appropriate technology movement, is opposed by many in the developing countries, who view it as another attempt by the developed countries to perpetuate existing patterns of domination.[19]

Certainly, the cynics in the developed and developing countries alike can point to the fact that there is little evidence to indicate that the developed world is eager to assist the South. In general, relatively small amounts of aid have been provided by the advanced countries and, despite all the talk, relatively little has been achieved, although there is little doubt that the South requires increased support if many of these countries are to meet even the basic needs of their people. In a new attempt to deal with this problem, an Independent Commission on International Development, chaired by Willy Brandt, was established to examine the entire question of North-South relations. In its unanimous report, whch was published in March 1980, the Commission emphasized the growing communality of interests between North and South in such areas as eliminatir g poverty, trade and commodity agreements, the development of natural resources, and the recycling of financial surpluses. The report specifically recommended the large scale transfer of resources to the developing countries, a strategy for international energy, a global food program, and the initiation of reform in the existing international economic system. Whether the work of this commission will prove more meaningful than the many other reports in this area that have been prepared in the past remains to be seen.[20]

Partly because of dissatisfaction with existing patterns of North-South flows, the possibility of increased cooperation among the developing countries themselves has received increased attention. The idea of sharing technological knowledge, experiences, and skills among them is an idea whose time may well have come. Until now, most linkages, be they commercial, intellectual, or technological, have been influenced by former patterns of imperialist control. Thus, former French colonies tended to look to Paris, former British colonies to London, and so forth. Conacts between and among countries in a region, even when they were neighbors, were very limited. Yet, the countries of the Third World share common interests and problems and, indeed, in recent years various international and regional groupings such as the group of 77 which seeks to change the rules of the present international economic order and the Andean pact have been created to increase their bargaining power and enhance the scope of cooperation.

The potential for cooperative activities in many areas has been well established, for a wide range of training and research institutions and numerous experts to provide technical assistance are available. In 1978 the United Nations sponsored a major global conference on Technological Cooperation among Developing Countries (TCDC) in Buenos Aires, in order to promote such cooperation and to identify bottlenecks that could be eliminated. One of these is the lack of information about the capabilities that are available in the Third World and, in order to deal with this constraint, the UNDP has initiated an information referral service.[21]

How successful this new approach will be in achieving higher degrees of mutual aid and self-reliance remains to be seen, but the potential is obvious. One should keep in mind, however, that the South contains a wide variety of states with very different resources, levels of development, orientations, and needs, and one can, therefore, envisage new patterns of dependency relationships with such countries as Brazil and India emerging as new "centers."

Moreover, the creation of such new alignments could have a profound impact upon the other relationships discussed above. Although these relationships have all been analyzed separately, it is obvious that the framework represents merely an organizing device for a complex set of problems, each of which has major spillovers; developments in one set influence more or less profoundly relations in another. If, for example, East-West relations were to improve markedly and agreements on military spending negotiated, additional resources would be available that could be diverted toward the developing countries. Similarly, decisions to help the South develop particular industrial capabilities may have important implications for East-West trade or for relations among the developed countries as they seek to accommodate the new production.

Conceptually, however, it appears that while technological issues permeate the major alignments within the present international system, technology *per se* has had little impact upon such cleavages. The critical variables remain, as V. Basiuk notes in Chapter 8, the political ones. In his words:

> . . . technology has not been successful so far in changing alignments of nation states or modifying the relationship between them where political considerations militate against such a development . . . The role of technology is likely to increase in the future as the spectrum of potential technologies and the cost of their implementation increasingly exceeds the resources of individual nations. However one should not overestimate the ability of this trend to overcome serious political and national security obstacles.

Technology has, however, had a direct and profound impact upon other elements of the international system, though in this regard too, political factors are often a major variable. This is particularly evident when one considers the variety of new "global problems" that have been created by the expansion of man's ability to exploit and control his environment and the difficulties of dealing with these problems. These difficulties are compounded by the characteristics of modern technologies that permit man to function in areas that were too inclement or out of his reach, such as seabeds, outerspace, and the polar regions, and permits nations to mine the oceans, modify the weather, and, perhaps, one day soon, colonize the planets.

Man possesses this ability without, at the same time, being able to anticipate the results of his actions for, as has been noted, technology produces second- and third-order consequences that are difficult to predict, yet which are often more significant than what is anticipated. Furthermore, even when potential hazards hve been identified, we often do not possess the knowledge that is necessary to make intelligent decisions. We seldom know where the critical physical threshholds are, or how far we can go before systemic changes occur. How many chlorofluorocarbons, for example, can be released into the atmosphere before the ozone layer is destroyed?

Moreover, these impacts may be delayed or occur slowly over long periods of time, seldom respect national boundaries, and often affect groups and areas that are distant from the initial application of the technology and from the decision-makers involved. Indeed, the application of modern technology increasingly requires operations that take place outside national boundaries and yet have profound impacts within them. The activities of multinational corporations, which I shall discuss below, exemplify this point.

Technological developments in transportation and communication have made this situation possible; one cannot, for example, conceive of multinational corporations without these. Essentially, the importance of space has diminished; a new era of mobility, both intellectual and physical, has opened. Recent decades have witnessed a remarkable increase in international exchanges of all kinds — social, cultural, political, and economic. Moreover, practically every developed and developing country has established goals that can be achieved only through greater participation in international affairs. Economic development requires exports and imports and, as was noted earlier, governments everywhere are increasingly involved in and affected by the flow of technology and other resources across national boundaries.

As a result, the distinction between domestic and foreign, between national and international, is rapidly losing its meaning. As I have already emphasized, it is essentially no longer possible to devise policies to deal with any major contemporary issue without considering the international implications of the policy. And, conversely, events in one part of the world have profound consequences for states far removed geographically.

These developments have spurred the emergence of new actors in the international arena. Traditionally, ministries of foreign affairs have not been concerned with technological problems, nor have they possessed the capability to deal with their policy implications in a sophisticated way. Efforts to provide these agencies with an enhanced scientific and technological capability have taken place in many countries, though with mixed results. In the United States, for example, a Bureau of Oceans, International Environmental and Scientific Affairs (OES) has been established in the Department of State, but the results have not been entirely satisfactory. Moreover, an increasing number of domestic agencies whose concerns have become internationalized are now actively involved in some aspect of foreign affairs. Hence, the Department of State, and its counterparts in other countries, often have to compete with organizations that possess far more developed scientific and technological competencies and resources. In the interbureaucratic battles that inevitably occur, these agencies often have a much greater voice in determining policy than the traditional ministry. The implications of this situation for United States technology policy are discussed by Wright in Chapter 4.

The new actors who have become involved in decision making on international issues within states have also entered the international arena. Scientists and technical experts of all sorts have become active on the international scene, as representatives of particular agencies of a state, as representatives of professional organizations of all kinds, and, in the case of the Pugwash movement, as members of the international scientific community.[22]

Scientists and their organizations represent but one category of the new actors who have emerged on the international scene. Perhaps the most powerful, at least the one which has received the most attention, is the multinational corporation (MNCs). MNCs have become a major force in international economic relations. They are complex organizations whose functioning is made possible by technology. An MNC is able to draw resources from its operations in different countries and to combine them in ways that enhance the organization's goals, to recruit and interchange personnel of many nationalities, and to function within and across national boundaries in a manner that effectively divorces it from identification with any particular nation-state. Moreover, the corporation's goals may not be consonant with those of a state in which its headquarters are located or where its units operate. Decisions are made on the basis of the best interests of the organization and subsidiaries are opened, expanded or closed, prices raised or lowered, R & D carried out in one location or another, with, often, little concern for the impacts that such decisions will have upon particular states. The MNCs' control over technologies and capital gives them considerable bargaining power and they are often able to extract concessions from weak states that they could not obtain elsewhere.

As a result of considerations such as these, measures to control MNCs, to establish international codes of conduct, and to enhance the bargaining capability of smaller states are being widely discussed and, in some cases, implemented. Nor is concern over the activities of MNCs confined to the developing countries. Many observers in the developed world are pointing to what are increasingly viewed as negative impacts upon Western economies and the United States will probably have to assess more carefully than heretofore whether it is desirable to permit MNCs to continue to operate as they have in the past. (See Leonard and Gilpin, Chapter 3.)

The number of international organizations has also increased greatly in recent years. Many of these have been created specifically to deal with issues created by technology, but it is difficult to find any international organization that is not involved with technology in some way or other. Essentially, they can be grouped into the following categories:

1. those (such as the World Health Organization and the Food and Agricultural Organization) that promote the development and diffusion of specific technologies,
2. those (such as the International Atomic Energy Agency, which administers safety regulations for civilian nuclear facilities) that regulate technologies,
3. those (such as the International Telecommunications Unions, which allocates radio frequencies) that develop and administer standards in technological areas, and
4. those (such as UNESCO) that promote the development of planning capabilities in science and technology.[23]

The proliferation of these organizations has not facilitated the ability of mankind to deal with the myriad of technologically related issues, even though questions of control and regulation at the global level have emerged as central issues of concern and even though, given the nature of the problems, especially their internationalization, such organizations may well be the most effective arena in which to devise and implement solutions to such issues.

Numerous reasons can be cited to explain why the existing structure of international organizations is inadequate including, ironically, their very increase which has aggravated problems of overlapping jurisdictions, inadequate funding, and administrative shortcomings. Indeed, the number of organizations is now so large that many countries with a limited pool of skilled manpower find it difficult to participate effectively in their work. Nor can one ignore the ways in which existing value systems and decision-making systems tend to emphasize short term costs for the groups involved. Hence, it is often extremely difficult to develop long range policies to deal with issues. Attempts to do so, to increase the time perspectives involved, are now common in many countries but, at the international level, little progress has been achieved in this regard to date.

Finally, one should note that the success or failure of any effort to resolve problems on a global level is profoundly influenced by the actions and attitudes of nation-states for, ironically, despite the myriad changes that have occurred, the nation-state remains the preeminent actor in the international arena, even though its power relative to other actors may decline, and the success or failure of multilateral attempts to resolve global problems depends on their commitments. Yet, most governments do not have much confidence in the ability of international organizations to deal with such issues, a situation to which, ironically, they contribute by their own policies and actions.[24]

Whatever the reasons, the contemporary world is characterized by weak institutional arrangements for the regulation of man's activities at the very time that the need for cooperation among states is accelerating. Technology has impacted upon the world in the manner discussed above; it has increased interactions, created global problems, blurred the boundary between domestic and foreign affairs. All of these consequences are commonly referred to as "interdependence," and talk of interdependence is so widespread that it is almost impossible to find a discussion of international affairs that does not refer in some manner to this phenomenon.

Despite its popularity, however, the concept does not accurately reflect the complexities of the new linkages and relationships that have been created among nations. In particular, it is important to note that not all states have been equally affected by these developments. Different states are more or less interdependent, for the integration that has occurred varies markedly. Some countries, such as the United States and Japan, are far more involved than Greenland or Zambia, for example. Two separate variables must be taken into account when considering this question. The first is the degree of involvement and participation, the second, the significance of these relationships for a specific state, that is, the vulnerability of a state to changes in existing patterns. Some states, including the United States, are, despite the rhetoric, far less dependent on the outside world than many others for, despite the commonly accepted belief that the welfare of the United States is based to a large degree on imports of natural resources, especially oil, the United States could pursue an isolationist policy based on the development of its own resources, though such a strategy would involve obvious costs. Nevertheless, the strategy is a feasible one whereas, as the "dependency theorists" have emphasized, much of the Third World is extremely vulnerable to decisions made in Washngton, London, Moscow, or Tokyo, both in terms of markets for their goods and in terms of the supply of technology and capital.

Yet, it is true that technological developments are constantly impacting upon the structure of the international system, that new linkages and relationships continue to proliferate and that new issues always arise. These issues are difficult enough to deal with within the context of the nation-state but, at the international level, the lack of consensus on most issues and the absence of supranational institutions that can establish and enforce particular patterns of behavior magnifies the difficulties.

There is no evidence that this situation will soon improve. Rather, it is obvious that the impact of technology will continue to create new problems and opportunities. The case of micro-electronic technology, which is discussed by Morehouse in Chapter 9, illustrates this point only too clearly. Micro-electronic technology holds the promise, and the threat, of producing a new industrial revolution with profound consequences for developed and developing countries, relations between and among them and many other dimensions of the international system. Indeed, this case highlights practically every issue which has been discussed in this chapter.

This technology is being developed with nation-states, but its development is spurred by the character of the international system, particularly patterns of international competitiveness. Because of their concern with maintaining the economic vitality of their economies, the United States, Japan, and other European countries are spending billions on R & D in this area. The diffusion of this technology will yield greater productivity and technological advances, but will also result in rising unemployment and shifts in comparative advantage and international competitiveness. How the particular costs and benefits will be distributed within and among different countries will depend upon the political decisions that are made and the kinds of policies that are implemented. This technology will also affect each of the major international relationships in numerous ways. The strategic balance between East and West may well tilt in favor of the United States and its allies if the United States can maintain a substantial lead over the Soviet Union in this area. Moreover, relations among the developed Western countries may also be transformed, for not all are equally advanced in this technology. Hence, the capabilities of some states, such as Japan, may be enhanced, while that of others, such as Sweden, adversely affected. As far as North-South relations are concerned, the gap between them may become ever larger as developed countries achieve new levels of technological capability that the rest of the world cannot match. And, if the developed countries are confronted with high unemployment rates, pressures for welfare and protectionist policies may prove irresistible, thus spurring increased interactions among the developing countries.

How micro-electronic technology will, in fact, impact upon the international system remains to be seen, for the future is not predetermined. As I stressed earlier, and as is evident from the analyses of the contributors to this volume, one should not assign a deterministic role to technology.

Technology defines the magnitude and direction of change that is possible, but other factors, notably political ones, profoundly influence the extent to which the potential is realized and the ways in which impacts are distributed among different groups within states, and upon states and groups of states.

This is perhaps the most important conclusion that can be drawn from this volume — there is no "iron law" of technology; mankind can determine its own future. The projections of the Club of Rome studies are no more accurate forecasts than the ones made by Jules Verne, Aldous Huxley, or George Orwell. Mankind will determine its own future, by design or by default. In the years ahead, it will have to cope not only with microelectronic technology, but with numerous other technologies as well. It is obvious that these will all yield major consequences, consequences that can be foreseen only dimly at present. What the result will be for mankind, whether the threats or the promises of these new technologies or some mixture thereof will be realized, whether human welfare will be promoted, will ultimately depend on man's ability to identify a vision of national and international social orders and to devise and implement policies that will create the kinds of institutional arrangements that can permit the realization of that vision.

Notes

1. The historical relationship between science, technology, and public policy is discussed by several contributors in I. Spiegel-Rösing and D. de Solla Price, eds., *Science, Technology and Society* (Beverly Hills, California, Sage Publications, 1977).

2. On the Ottoman case, see Bernard Lewis, *Emergence of Modern Turkey* (London: Oxford University Press, 1960); on Japan, see Noel Perrin, *Giving Up the Gun* (N.Y.: David Garvine, 1980).

3. For the present "state of the art," see Spiegel-Rösing and de Solla Price, *passim*. See also W. Kintner and H. Sicherman, *Technology and International Politics* (Lexington, Mass.: D.C. Heath, 1975), pp. 34 ff.

4. See Langdom Winner, *Autonomous Technology* Boston, Mass: MIT Press, 1979), pp. 8 ff., for a good discussion of the evolution of the concept and its present meaning. The definitions are taken from Bernard Gendron, *Technology and the Human Condition* (New York: St. Martin's Press, 1977), pp. 22–23; and David M. Freeman, *Technology and Society: Issues and Assessment, Conflict and Choice* (Chicago: Markham, 1974) pp. 5–9. See also E.G. Mesthene, *Technological Change: Its Impact on Man and Society* (New York: Mentor Books, 1970).

5. John V. Granger, *Technology and International Relations* (San Francisco, W.H. Freeman and Co., 1979) p. 11.

6. There is a rich literature on this subject. For a good introduction, see C. Freeman, "Economics of Research and Development" and E. Layton, "Conditions of Technological Development" in Spiegel-Rösing and de Solla Price, *op. cit.*, chapters 6 and 7.

7. This discussion and the data are based on Colin Norman, *Knowledge and Power: The Global Research and Development Budget* (Washington, D.C.: Worldwatch Institute Paper Number 31, July 1979).

8. The discussion that follows is based on the typology presented in *An Introduction to Policy Analysis in Science and Technology* (Paris: UNESCO, Science Policy Studies and Documents No. 46, 1979).

9. E.M. Graham, "Technological Innovation, The Technology Gap and U.S. Welfare," *Public Policy*, Spring 1979, p. 201.

10. See, for example, *Industrial Innovation and Public Policy Options* (Washington, D.C.: National Academy of Engineering, 1980).

11. F. Sagasti, *Science and Technology for Development: Main Comparative Report of the STPI Project* (Ottawa, Canada: International Development Research Center, 1978).

12. On this issue, see S.A. Lakoff, "Scientists, Technologists, and Political Power," in Spiegel-Rösing and de Solla Price, *op. cit.,* ch. 10.

13. For this discussion, I have drawn on my "The Prospects for Scientific and Technological Development in Saudi Arabia," *International Journal of Middle East Studies,* vol. 10, 1979, especially pp. 356 ff. Detailed references are contained therein.

14. On this issue, see the fine analysis by R.R. Nelson, "Less Developed Countries — Technology Transfer and Adaptation: The Role of the Indigenous Science Community," *Economic Development and Cultural Change,* October 1974, pp. 61–76.

15. See, for example, Ziauddin Sardar, *Science, Technology and Development in the Muslim World* (London: Croon-Helm, 1977).

16. On this issue, see the analysis by J. Tumin, "On East-West Trade," *New York Times,* October 19, 1980.

17. Granger, *op. cit.,* pp. 78 ff.

18. Tumin, *loc cit.*

19. For a sophisticated analysis of these paradigms within the Latin American context, see E. Haas, "Three Strategies for Managing Technology," unpublished manuscript. The second paradigm is reflected in several of the articles, especially those by Galtung and Morehouse in W. Morehouse, ed. *Science, Technology, and the Social Order* (N.J.: Transaction Books, 1979). For the third paradigm, see J. Ramesh and C. Weiss, eds., *Mobilizing Technology for World Development* (N.Y.: Praeger Special Studies, 1979).

20. The report was published as *North-South — A Program for Survial.* A summary is contained in *World Bank Annual Report, 1980.*

21. Norman, *op. cit.,* pp. 41 ff., "Bridges Across the South," *Development Forum,* July 1978, p. 4.

22. See Schroeder-Gudehus, "Science, Technology and Foreign Policy" in Spiegel-Rösing and de Solla Price, *op. cit.,* pp. 490 ff.

23. Granger, *op. cit.,* pp. 42 ff.

24. E. B. Skolnikoff, *The International Imperatives of Technology* (University of California, Berkeley, Institute of International Studies, 1972), p. 494.

PART ONE

Technology, And Developed Societies

2

Toward a Theory
of the Modern State
Administrative Versus Scientific State

JURGEN SCHMANDT

In this paper I shall discuss the concept and characteristics of the scientific state. I use the term for a system of governance that is shaped, to a significant extent, by science and technology. New public policy tasks, new forms of political organization, and new ways of policy making are involved. They add up to a new form of the state. In discussing the new developments I shall deal primarily with policy making and administration at the national level. But regional and local governments are undergoing similar changes, though often with some delays. International organizations are equally affected. I shall restrict myself to changes that are structural and systemic, leaving aside specific technological developments, such as modern weapons and computer systems. They have been the vehicles for many of the second-order changes that I will concentrate on. I shall say little, in addition, about changes that have not yet taken place. Nor will I address normative questions. Whether the development of the scientific state is desirable, dangerous, or both, depending on particular circumstances, is a critically important question. But attempts to find satisfactory answers have been stalled, since the debate about science and the state got caught in a sterile polarization between Orwell's specter of technocracy and Bacon's millennium of the scientifically administered polis. In this situation, it will be useful to concentrate on concrete changes and to ask what they add up to in terms of a general theory of the state. Once this task is further advanced, the philosophical debate will have a better chance to take a fresh look at the relationship between values and the new reality of the scientific state.

To gain historical perspective, I shall draw distinctions between the scientific and the administrative state. The theory of the administrative state was developed by Max Weber. His work will be discussed first. I shall then review some more recent writings on the development of the administrative state. This will help to see the administrative state as the response of the public sector to the policy tasks that resulted from industrialization of the economy, urbanization of the rapidly growing economy, and development of a sophisticated technical infrastructure in transportation, communications, and energy. These massive changes in economy and society brought with them unprecedented increases in the volume and diversity of administrative tasks and the need to provide an array of physical and social services. In its fully developed form the administrative state serves the needs of the welfare state. I shall then turn to a number of technological and societal changes that resulted in a new set of challenges to the public sector, among them scientific warfare, an innovation-intensive economy, and man-made threats to health and the environment. The scientific state will be introduced as the public sector response to the needs of a society that is dependent on scientific knowledge, and the technologies derived from it, and tries to cope with the complexities that have been created by these very factors.

My central theme is that knowledge is essential for both the administrative and the scientific state. The former depends on legal knowledge and methods. The latter continues to make use of this base but becomes equally dependent on scientific information and methods. Thus, the scientific state does not replace the administrative state, but meshes two modes of thinking, information processing, and problem solving. The search for a workable combination of legal and scientific rationalities is at the heart of institutional and procedural changes in government. Both forms of rationality need to be understood. Legal rationality will be discussed in the context of Weber's work. Scientific rationality, in terms of its political implications, manifests itself in three distinctive forms. I call these modes of the scientific state "science as product," "science as evidence," and "science as method." *Science as product* refers to the role of science in the process of innovation and the attendant functions of government in promoting and controlling science, as well as the technologies based upon it. The goal of government involvement is not to produce knowledge as such, but to make use of it in socially desirable ways. *Science as evidence* deals with the need, in order to make informed decisions, to base political solutions on the findings of empirical research about the nature of the problems at hand. Government would be blind without reliance on scientific evidence. *Science as method* means the use in government of policy analysis as an aid to decision making.

An increasing part of government activities uses working methods that are derived from science. I shall use these three different methods of the interface between science and government to structure my discussion of the scientific state and to illustrate how science has permeated the entire system of government.

This study, then, looks for answers to these questions: What is the nature of the scientific state? How does it compare with the administrative state? Is the transition from one state to the other in the form of adjustments and reforms, or does it represent a more radical change? To what extent is the development of the scientific state a global phenomenon, or is it limited to some nations, such as those large and affluent enough to support a full-fledged science and technology enterprise? Are smaller and less developed nations, because of resource constraints, limited to building partial copies of the master model developed by the most advanced world powers, or do different roads lead to the scientific state?

Weber's Theory of the Administrative State

Max Weber developed his theory of the administrative state around the turn of the century. His work is widely recognized as an insightful conceptualization of the modern state and its bureaucratic organization. Weber's political sociology has been influential in the United States. Talcott Parsons and Reinhard Bendix, in particular, have been instrumental in transmitting Weber's concepts and ideas (Parsons, 1937; Bendix, 1962). Even today, much of Weber's work has not been translated into English. But a three-volume edition of his *Economy and Society* — the central work in the present context — was published in 1968. Bendix's "intellectual portrait" of Weber combines carefully selected and faithfully translated excerpts from his major writings with analysis, placing Weber's thoughts in the context of German scholarship of the time and identifying central concepts and lasting findings.

Historical research, undertaken long after Weber's death, has provided detailed evidence that the state characterized by functional specialization and hierarchical decision making by trained public servants is a modern concept. Early stages of development are found at the end of the Middle Ages and during the Renaissance (Brunner, 1943). The administrative state, making use of large bureaucracies, is hardly a century old. Weber, using the legal and historical sources available to him, had come to similar conclusions. He distinguished among three forms of governance or, using his preferred term, of political domination — traditional, charismatic, and legal. Legal domination led to the development of the administrative state. Elements of this process can be found throughout history, but it was fully developed, according to Weber, only in the Occident and was closely related to the reception of

Roman law and the emergence of legal scholarship in the universities of the late middle ages (Bendix, 1962: 414, 418).

In the present context, only legal domination is of interest, since it represents the principle on which the modern state is built. Weber's concepts of traditional and charismatic leadership need be explained only briefly. Traditional domination accepts the authority that has "always" existed. The master reigns in accord with long-established customs and exercises his rule through household officials or vassals. Patrimonial and feudal systems represent this kind of domination. Charismatic domination, by contrast, is centered around an all powerful leader — prophet, hero, or demagogue. He is surrounded by a group of devoted followers who exercise official functions in recognition of their devotion to the leader, with little regard to specialized training and objective qualifications. Throughout history, the three forms of political domination are always found in combination. They do not exist as "pure types," except as theoretical constructs for purposes of analysis (Bendix, 1962: 296).

The basic principle of legal domination is obedience to the statute, rather than to a person. All persons are subject to the law and are equal before the law. These concepts are alien to the other forms of political domination and have emerged only as the result of a long historical process. Legal revelation and imposition of law by secular and theocratic powers represent two stages of this process. Lawmaking by legal professionals represents the third stage, which alone leads to the modern state. The modern state has three characteristics:

1. the administrative and legal order is subject to change by a formalized process of legislation;
2. the administrative apparatus conducts official business in accordance with existing legislation;
3. the state exercises binding authority over all persons in the area of its jurisdiction;
4. the state can use force in accordance with enacted statute. "Legal order, bureaucracy, compulsory jurisdiction over a territory and monopolization of the legitimate use of force are the essential characteristics of the modern state." (Bendix, 1962: 418.)

The legitimacy of the statute, under this system of governance, is based on the belief that the formalized process of lawmaking is just and equitable. If this faith is lost, the rule of law will have been eroded. The political community is not based on belief in some substantive value or purpose, but on belief in the process of legal rationality. The statute is binding, since it results from a rational process of legal analysis and formal procedures of formulation and adoption. In Weber's words: "The basic conception is: that any legal norm can be created or changed by a procedurally correct enactment."

(Quoted in Bendix, 1962: 419.) This definition is basic for Weber's theory of the administrative state. It raises two questions: Is this formalism void of ethical substance? What is the nature of legal rationality, which legitimates the entire political system?

Legitimation of statutes through formal procedures of enactment has been criticized as empty formalism which removes the political process from its normative and constitutional foundations (Mommsen, 1959). This interpretation places Weber among the intellectual fathers of Nazi ideology, since "even the most pernicious laws would be compatible with a legitimate system of legal domination." (Bendix, 1962: 484.) But this misrepresents Weber's thoughts. He does not ignore the role of substantive values upon which legal domination must be built. But Weber's sociological perspective compels him to state that natural law and basic values vary among nations and change over time. They reflect class preferences, and are eventually superseded by different values (Bendix, 1926: 419). The viability of the system of legal domination depends on its ability to find and maintain a balance between formal and substantive elements of the law. This balance must be constantly reworked in the light of changing circumstances and ideas. For the modern Western state, the balance is embodied in the rights of the individual and the supreme power of the people. Weber puts it this way: "The State is not allowed to interfere with life, liberty, or property without the consent of the people or their duly elected representatives. Hence any law in the substantive sense must . . . have its basis in an act of the legislature." (Quoted in Bendix, 1962: 422.) The formal process of legislating, by its very rationality, is expected to ensure recognition of substantive values, whatever these may be for a given society and time. Legal systems that no longer derive their legitimacy from religious revelation or tradition, are based on natural law. Their norms are binding because of their imminent qualities of logic and rationality (Weber, 1976: 497). The political philosophers of the seventeenth and eighteenth centuries in Europe, and the reflection of their concepts in the basic documents of the French and American revolutions, represent the most noteworthy examples of the impact of natural law on specific legislation. The rational law that results from this process contrasts with revealed or empirical law, since it relies on a few logically coherent substantive principles. Using these principles as a point of departure, the details of the legal system are developed through systematic elaboration of statutes and professional administration of statutes by persons who have received their legal training in a learned and formally logical manner (Bendix, 1962: 391).

Weber's concept of legal rationality is directly tied to the tasks and training of the legal profession. Their special skill is legal rationality, defined as "the capacity to state clearly and unambiguously the legal issues involved in a complicated situation." (Weber in Bendix, 1962: 414.) The importance of legal training for the rational approach to administrative tasks reflects the conditions of German judicial and administrative history. Weber himself was trained as a lawyer and was deeply influenced by the legal scholarship of his time. In his research, he branched out into economics, sociology, and political science. But the dominance of legal concepts and methods is evident in his analysis of the political process. This can be illustrated by his view of the superiority of the continental method of legal reasoning over the common law approach in England.

Training in the common law was dispensed on the job, or in guild schools. Apprenticeship with a master was the rule for attorneys-solicitors who represent the client. Formal training in so-called Inns of Court was dispensed to advocates or barristers who represent a case in court and virtually lose direct contact with the client. But even for this group, training remained entirely practical, always moving "from the particular to the particular," never "from the particular to general propositions in order to be able subsequently to deduce from them the norms for new particular cases." (Weber in Bendix, 1962: 412.) By contrast, legal instruction in the continental universities, following the reception of Roman law, emphasized the logical structure of this body of law, while discarding many of its particular provisions. The goal was to search for abstract principles on which specific rules would be based (Weber in Bendix, 1962: 415). Under this approach, legal construction of general principles was the most important task of jurisprudence, and became the model for the discharge of judicial and administrative tasks. The reception of Roman law brought with it the establishment of a new legal elite whose members proclaimed "the principle of restricting legal reality to those concepts and rules that could be thought and constructed by the legal expert." (Weber, 1976: 492.)

Legal construction are the key words here. It was a central issue in legal scholarship around the turn of the century. Rudolph Sohm, a leading jurist of the period, whom Weber quotes on occasion, defined the central task of legal scholarship as "Interpreting the law . . . shaping today's law out of yesterday's laws." (Sohm, 1911: 36.) Legal principles needed to be abstracted inductively from particular cases. Subsequently, they were applied to analogous situations. Rudolf Ihering, writing at the time of Weber's birth, identified analysis, concentration, and construction as the major steps in developing legal concepts. Construction was the most difficult task. It consisted of using the legal principles derived from analysis and concentration to formulate new law. This creative process required logical reasoning, professional training, and objective interpretation of legal sources (Wolf, 1939: 503). Weber used a similar set of concepts to define legal rationality (Weber, 1976: 395-396).

Legal rationality, as defined by Weber, developed at a particular time in history, beginning with the reception of Roman law during the Renaissance and culminating with the scholarly approach to legal construction in nineteenth century German jurisprudence. The administrative state as a distinct system of governance is intimately linked to these historical developments. However, the characteristics of the administrative state proved to be general qualities that were not limited to the reach of the continental legal system. This was the case because the imperatives of industrialized economies encouraged the development of different, though equally rational, systems of law in the Anglo-Saxon countries. Thus, legal rationality became the underlying principle of the modern state everywhere. Speaking about the emergence of comparable political structures in Europe and in the Anglo-Saxon countries, Bendix remarks: "Although the development of the law in the two cases was strongly influenced by law specialists of different types, both legal systems were compatible with the increasing political and economic interest in a dependable legal system." (Bendix, 1962: 416.) What Weber first identified as legal rationality in analyzing successive attempts to learn from Roman law, led him to the discovery of the basic structures of the state of the industrial era. I shall now turn to a discussion of how the principle of legal rationality, according to Weber, determines important organizational and procedural characteristics of the administrative state.

Legal rationality is translated into administrative practice with the help of hierarchically organized bureaucracies which conduct official business in accordance with strict procedural rules. "Bureaucracy is the technically purest type of legal domination." (Weber, 1956: 153.) It alone is capable of dealing with the scope and size of administrative tasks in the modern state. "For the need of large-scale administration today bureaucracy is inescapable." (Weber, 1976: 128.) The only alternative is between bureaucratic and dilettante forms of administration. Bureaucracy owes its superiority to a number of innovations which developed slowly over time and made possible the modern state. Taken together, they are the vehicles for extending legal rationality into administrative rationality.

Weber lists six innovations: (1) State business is conducted on a continuous basis and in functionally specialized offices. At first, functional specialization is limited to the administration of justice, taxes, and military matters; other functions are added over time. (2) Administrative decisions are made according to rules. Where necessary, trained professionals will exercise informed judgement in the context of established rules. (3) Business is conducted on the basis of written documents. (4) Offices are organized in hierarchical layers with different levels of supervision, as well as the right of appeal to higher authority. (5) Officials are accountable for the execution of assigned duties. (6) Officials do not "own" or inherit the offices they occupy, but are appointed according to objective criteria on the basis of qualifications and merit (Bendix, 1962: 424).

Bureaucracy is superior to other forms of administration because it embodies knowledge. "Technical knowledge is required for the discharge of administrative tasks due to the modern technology and economy of production, and it is irrelevant whether production is organized in a capitalistic or socialist fashion." (Weber, 1976: 128.) Weber sees an explicit connection between the administrative state and the complexity of the modern technology-dependent economy. He also states that different political systems are equally dependent on specialized bureaucratic structures staffed by highly trained public officials. To a large extent, the administrative state is determined by the technological and economic realities of the industrial age to which it attempts to respond with informed political and administrative decisions.

As in the case of the other characteristics of the administrative state, the development of a trained class of civil servants was a gradual process. Increasingly, paid full-time administrators replaced unpaid avocational notables who had discharged certain administrative tasks. Historically, the new class of civil servants developed first in the service of the absolute state in France and Prussia. Barker characterizes them as those political systems that "made a science of the service of the state"; England, by contrast, "considered it a task for intelligent amateurs." (Barker, 1966: 29.) States differed in the ways civil servants were trained. The French, for example, placed great emphasis on providing training that combined technical competence with administrative skills. In Germany and Scandinavia, legal training became dominant around the end of the eighteenth century, and senior civil service positions were almost entirely reserved for law graduates in Weber's time. A different orientation characterized training of aspiring civil servants during the early stages of professionalization. In Prussia and Austria, for example, administrative skills were emphasized that were needed to manage both the state and the personal dominion of the monarch — two tasks that were not yet fully separated from each other. This was the system of cameralism under which training "included predominantly what we would today refer to as public finance, including both revenue and expenditure administration, police science, and economics, with particular emphasis upon agriculture." (Mosher, 1968: 33.) Mosher adds an interesting observation. "In some ways, cameralism was the principal precursor of the development of public administration in the United States in the first half of the present century." (Mosher, 1968: 33.) American students of cameralism, upon returning from Europe, shaped the new science of public administration around the modern-day subject matters of political economy, administrative theory, political theory, and the beginnings of the social sciences in general (Friedrich, 1939).

For Weber, the civil servant, as he knew him in his time, was an indispensable instrument of government. "The entire history of the modern state is identical with the development of modern civil service and of bureaucratic administration." (Weber, 1956: 154.) The civil servant has two outstanding characteristics: He commands knowledge essential for the task at hand and he is motivated by the ethos of serving the state and the citizenry.

Weber distinguishes between two forms of knowledge that are essential for the public servant. These are "discipinary knowledge" and "service-related knowledge." The first form is primarily legal knowledge, but also includes specialized knowledge in such areas as public health, engineering, or military matters. However, Weber has little to say about the non-legal knowledge requirements of modern administration beyond acknowledging that it is needed. Service-related knowledge "is acquired in the performance of agency tasks" and takes the form of "factual information documented in agency files." (Weber, 1976: 129.) Both forms of knowledge tend to reinforce each other. "The bureacuracy is domination through knowledge. It is powerful through the use of disciplinary knowledge and further strengthens its power base by the use of service-related knowledge." (Weber, 1976: 129.) The modern civil servant joins the ranks of other professionals — judges, physicians, scientists, managers, engineers — who alone can apply a body of knowledge to the provision of services critical to the functioning of society. "The modern civil servant, reflecting the rational and technological nature of modern society, must be increasingly trained and specialized." (Weber, 1966: 47.) He functions as an expert, and the expert, not the cultivated man, is the educational ideal of the bureaucratic age (Bendix, 1962: 430).

The ethical dimension of public service, in Weber's view, is an important condition for making the bureaucratic system work. The duty of the public servant is to act according to the statutes and to exclude all personal considerations. At one point Weber calls this principle of public administration "the domination of the formalistic nonperson." He seems to have recognized both strength and weakness in this approach. But he clearly was most impressed with the improvement in public service by working strictly according to the rules. "The ideal civil servant performs his duty without hatred and passion, without love and enthusiasm, only doing his duty. He treats all persons in a similar situation in exactly the same way, without regard to individual status." (Weber, 1976: 129.)

The civil servant, then, is a powerful and knowledgeable professional and an influential member of the ruling elite. "The civil service has shown its excellence in the pursuit of public, well-defined functions through its sense of duty, its objectivity, and its aptitude for handling organizational problems." (Weber, 1966: 76.) The high social rank accorded to the civil servant closely reflects the situation in Germany around 1900. Senior public officials were considered to hold positions similar to influence and social standing to those held by judges. A British author, comparing different administrative systems in Europe, describes the German ideal in words that could have been Weber's: The Public servant's ". . . primary task is the application of the law, dealing objectively with those matters which arise in the course of administration, finding the law appropriate to the particular case, and making a decision entirely within the terms of that law. . . . Legal knowledge and the qualities of impartiality and disinterestedness are clearly needed for both administrator and judge." (Chapman, 1966: 100-101.) As public functions become larger and more diversified, the power of the public servant increases and makes the functioning of the state dependent upon him.

This must not mean, however, that he dominates the political system. Weber is explicit about the dangers of such a condition. Not only are there historical illustrations for the usurpation of the political system by the bureaucracy — in Weber's own time, evidenced by the inflexible and elitist attitudes adopted by many officials in the imperial administration of Germany — but the very success of bureaucratic administration encourages this trend. The bureaucracy's command over technical information makes it a superior tool of administration. If access to information, and knowledge how to use it, is denied to the political leaders, their control functions are eroded. Political control not only requires authority to supervise, but also knowledge to do so effectively (Bendix, 1962: 451). Monopolization of knowledge, on the part of the administration, leads to bureaucratic absolutism — a serious and constant threat to the rule of law.

Bureaucracy has a tendency to keep information about its work tightly guarded. Rules to protect the confidentiality of official business are appropriate when the outcome of intended action depends on a contest with some hostile organization. But the tendency to keep information secret, and thus to avoid outside inspection and control, is a powerful motive of administrative behavior in general. In Weber's discussion of the "pathology" of legal domination, he is genuinely concerned with the threat to effective political control in regard to the management of information. The problem has two aspects. The administration will attempt to deny access to information. In addition, the political authorities responsible for control of administrative action may not know enough to make intelligent use of information.

Different forms of political control are discussed. For each department or ministry of the government, political leadership must be exercised by an appointed or elected political official. He is not a civil servant, or is so only in a formal sense. Given his political status, it is correct not to require of him the technical training that the civil servant must have to occupy his position. Nevertheless, he must not be a dilettante or parvenu. He will know less about the technicalities of the job, but he must be able to act intelligently on the basis of information presented to him by his agency staff. His most important function — one that makes him different from the civil servants working under him — is political in nature. He has to define the major goals of policy making and fight for them, with *ira et studium* — the exact opposite of the civil servant. The same applies to the leader of the government as a whole, for whom fighting for power and political objectives is the most important task. In addition, he is responsible for control of the administration by insisting that the government's decisions are carried out (Weber, 1976: 836-837).

The highest authority to exercise political control lies with the representative assembly. (Weber does not mention the courts as an instrument of ultimate political control. Once again this reflects the constitutional condition in the Germany of his time.) Efficiency of control, in the case of the legislature, is most heavily dependent on access to administrative information. "Only access to official information, independent of the good will of the civil servant, guarantees effective control of the administration." (Weber, 1966: 77.) Parliament, to be effective, must have the right to subpoena administrative witnesses and documents. The German *Reichstag* did not have this right until the Weimar constitution came into being at the very end of Weber's life. Weber compared the failure of the parliamentary system in Germany to the British practice where the work of parliamentary commissions "represents the best achievements of the English parliament." (Weber, 1976: 855.) He concludes that only the right of legislative oversight, with full access to administrative information, makes the parliament into an effective political institution (Weber, 1976: 856). Legal domination, as the superior form of political organization, depends on bureaucracy staffed by highly trained and dedicated professionals who are subjected to political control on the part of agency heads, government leaders, and the legislature. The flow of information among the elements of the political system determines its quality. A sound system of legal domination uses and generates a great deal of information that will be available, for its most important parts, to the political authorities responsible for providing leadership and control.

This is Weber's attempt to reconcile the need for efficient and strong administration with the even higher need for the exercise of political control and respect for the will of the people. He does not see the existence of a large state machinery as being incompatible with a democratic system of government. Nor does he believe that they are easily combined with one another. To achieve the right balance requires constant effort and vigilance. The administration must not become an end in itself. Only political leadership and parliamentary control can prevent the administrative function from becoming predominant. Once political institutions fail in this respect, bureaucracy has become dysfunctional. It is then either administrative tyranny, or technocracy, or dull regimentation, or some combination of these various forms. All are dehumanizing and eliminate political freedom and participation. Since Weber's thoughts have often been misrepresented as giving undue importance to the adminsitrative function, it is important to recognize that he saw it as a necessary but dependent force of a healthy political system. Once it becomes independent, it damages the state. Bendix sums up Weber's position as follows:

> Weber distinguished leaders who have the power to command from officials who stand ready to execute the orders they receive, because these functions involve a fundamental difference in responsibility. . . . He clearly did not maintain that politicians and administrators behaved in this way. He contended, rather, that the abdication of leadership-responsibility by politicians and the usurpation of policy-functions by administrators are ever-present hazards of government under the rule of law. (Bendix, 1962: 485-486.)

Post-Weberian Views of the Administrative State

Weber's central concept is legal rationality, which distinguishes the modern state from earlier forms of political organization. Administrative rationality is a subsidiary concept, since administration, as we have seen, is subject to political guidance and control. By contrast, most of the literature on Weber's theory of the state deals with the characteristics and dangers of bureaucratic administration without giving much attention to the checks and balances that are an integral part of Weber's view of the system of legal domination. With hindsight, Weber can be faulted for not giving sufficient attention in his analysis to such issues as public participation and the details of managing knowledge generation and utilization in government. He certainly underestimated, in addition, dangers of the administrative state resulting from a highly formalized and depersonalized concept of official behavior. It is wrong, however, to interpret Weber as an advocate of bureaucratic dominance.

Such interpretations can be found in surprisingly large numbers. Bendix comments on this tendency among German writers, such as Wolfgang Mommsen and Christian von Ferber: "After all, we would not assess Freud's theoretical contributions, either, if we denigrated them because they helped to foster the 'failure of nerve' that characterizes the modern climate of opinion." (Bendix, 1962: 417-472.) An example of this misrepresentation in American literature is to be found in Vincent Ostrom's work. According to him, Weber identifies bureaucratic organization with a political system based upon a "monocratic" principle. Only this form of administration is efficient. Democratic administration, by contrast, is unworkable (Ostrom, 1973: 78). Ostrom does not seem to have used recent Weber translations, in particular the 1968 edition of *Economy and Society*. His references are all to the outdated Rheinstein (1954) selection. Weber, as we have seen, draws a distinction between bureaucracy in a system of legal domination and "bureaucratic absolutism." The latter, according to Weber, is bureaucracy encroaching upon the political process. Ostrum seems to suggest that this pathological form of bureaucracy represents the healthy norm for Weber. Other authors present the same one-sided view of Weber's theory (Henry, 1975: 573).

Michael Crozier, in his study *The Bureaucratic Phenomenon,* points to the almost schizophrenic nature of much of the literature dealing with bureaucracy and, directly or indirectly, with Weber:

> During the last fifty years, many first-rate social scientists have thought of bureaucracy as one of the key questions of modern sociology and modern political science. Yet the discussion about bureaucracy is still, to a large extent, the domain of myths and pathos of ideology. On the one hand, most authors consider the bureaucratic organization to be the embodiment of rationality in the modern world, and as such to be intrinsically superior to all other forms of human organization. On the other hand, many authors — often the same ones — consider it a sort of Leviathan, preparing the enslavement of the human race. This paradoxical view of bureaucracy in Western thought has paralyzed positive thinking . . . and has favored the making of catastrophic prognostications. (Crozier, 1964.)

Crozier recognizes that Weber's concept of bureaucracy is built around an important insight about the conditions of organizational success. The principle embodied in bureaucratic organization is the "rationalization of collective activities (which) is brought about by concentration of orgniza- tions and, within these, of a system of impersonal rules." (Crozier, 1964: 7.) However, the reality of much of real-world bureaucratic behavior is worlds apart from the ideal type. Why is this so? This is the question that Crozier tries to answer in his study of large public sector organizations in France. His focus is on a wide range of dysfunctional aspects of bureaucratic behavior, which have become more pronounced and dangerous for the functioning of the public sector as organizations have grown and become responsible for numerous tasks that directly affect people's lives.

Crozier analyzed such well-known attributes of bureaucracy as slow- ness, ponderous style, unnecessary complexity of rules and procedures, pre- occupation with performance of routine tasks, loss of organizational flex- ibility and innovation, lacking attention to policy concerns, weak leadership, inability to change, or lack of response to the needs of clients. All of them are part of a common syndrome, and manifestations of an underlying disease. A common explanation is to blame the loss of purpose and responsiveness on the tendency of large organizations to become dominated by technical con- cerns. This gives the agency a machine-like quality that is resented by the workers and offensive to the clients. But, rather than seeing the development of bureaucratic dysfunctions as the result of some technological imperative,

Weber's analysis of the administrative state suggests a more man-centered explanation. If political guidance and control are weak or missing, the observed pathologies emerge less as a result of technological factors, but more as the human response to loss of purpose and motivation, which is compensated by an unwarranted concentration on technical aspects of agency tasks and a tendency to perceive these as all important in their own right. Lack of political control was indeed what Crozier found at the root of the bureaucratic malfunctions in the agencies studied by him. He used his findings to discuss ways in which bureaucracies can be revived.

Better understanding of organizational behavior, and a recognition of its role in the public sector, make it possible to improve the linkage between bureaucratic and political functions. Progress in the techniques of management makes agencies more flexible. People have grown more sophisticated in dealing with complex social environments. This makes them better prepared to deal more intelligently with large bureaucracies. These various trends, taken together, tend to make the concentration of administrative power less oppressive. "Organizational progress has made it possible to be more tolerant of the personal needs and idiosyncracies of individual members: one can obtain the wanted results from them without having to control their behavior so narrowly as before. Cultural sophistication, on the other hand, has increased the individual's capacity for accommodation and . . . his possibilities of independence." (Crozier, 1964: 292.) Thus, the real potential of bureaucratic organization, which had been endangered by its pathologies, will re-emerge. "Elimination of the 'bureaucratic systems of organization' in the dysfunctional sense is the condition for the growth of 'bureaucratization' in the Weberian sense." (Crozier, 1964: 299.)

This brings us back, in reflecting upon the role of the contemporary state, to the central theme in Weber's analysis — the relationship between administrative rationality and political leadership. This, and not the question of technocracy, is the most critical issue, and the role of knowledge and expertise needs to be placed in this context. Crozier, in discussing the danger of technocratic rule on the part of administrators, makes an important observation. Technocracy, as the rule of politically unauthorized and unaccountable experts, can only take over in a political void. An active and well-informed leadership will not tolerate this usurpation of power. "When progress accelerates, the power of the expert is diminished and managerial power becomes more and more a political and judicial power rather than a technical one. Managers' success depends on their human qualities as leaders and not on their scientific know-how." (Crozier, 1964: 300.) Technocracy is the result of administrators adopting a narrow and technical view of their responsibilities and of political leaders failing to provide leadership. It is not just the work of experts groping for power. Obviously, this may be begging the question, since nothing is said about how political leadership is to be exercised in a knowledge-dependent political environment. But this question will have to be postponed until the discussion is focused explicitly on the nature of the scientific state.

At this point, I shall use Redford's *Democracy in the Administrative State* in order to identify more precisely the requirements for workable linkages between administrative and political functions, and institutions in the contemporary state. This discussion should also be helpful for a consideration of the role of the citizen which, in Weber's theory, is limited to the important but passive status of equality before the law.

Redford begins with a definition of the administrative state, which, though not directly based on Weber, is Weberian in spirit. "This is a political-administrative system which focuses its controls and renders its services through administrative structures but includes also the interaction of political structures through which these are sustained, directed, and limited." (Redford, 1969: 4.) Redford, along with virtually all contemporary scholars in political science and public administration, declares as unworkable and artificial any attempt to impose a strict division of responsibilities between policy making and administration. Both functions are intertwined. Yet, policy and administration meet with each other in different ways, and with different proportionate roles, in different parts of the political system. For the American system of government, Redford distinguishes among three levels, each of which is characterized by a unique mix of policy making and administration, and each leads to different interactions among government, interested parties, and citizens. These relations determine the quality of the political system — its openness, responsiveness, efficiency, and the degree to which it can qualify as democratic (Redford, 1969: 69). If contacts are closed and secretive, few in number, and without much impact on decisions, the political system, whatever its official constitution, is undemocratic, since it is incapable of responding to leadership and citizen participation. These institutional and group relationships, which are different for each level of government, are similar to Weber's concept in so far as they focus on the process of policy making. They are the conduits for what Weber calls legal rationality. They go beyond Weber in so far as they allow for different forms of legal rationality, according to a variety of political tasks. This diversification provides for increased flexibility of the overall system and multiple access points for citizen and interest group involvement.

Redford calls the highest level of the political system the macro-level of policy making. Decisions made at this level are large, strategic, and likely to be hotly debated. Presidential and Congressional policy making is involved. Substantively, "basic decisions on rules for society and roles for actors in the administrative state are made." (Redford, 1969: 69.) Actions taken, as well as procedures used, are highly political. Consequently, on the policy-administration continuum activities at the macro-level are placed predominantly on the left side of this line. There is little room for "submerging politics in administration," for example, "through prescription of rules and roles, and delegation to professionals." (Redford, 1969: 114.)

The first level down, action is entrusted to the subsystem or program level of policy making. At this level, government is functionally diversified. Interactions are intense among agencies and bureaus, their counterpart congressional committees and subcommittees, and interest groups. The trade press and lobbyists concerned with a particular area of program specialization also are heavily involved. Political subsystems are accessible to many interests, though perhaps in a limited way in their early stages of formation. But, the policy environment of the subsystem is largely confined to itself. "Substantial changes in the balances among interests served by subsystems can be expected to occur only through macropolitical intervention that modifies the rules and roles operating in the systems." (Redford, 1969: 105.) The large institutional network at this intermediary level of the political system performs many of the functional tasks of the administration state. The agencies and officials serving here provide many technical services required of the modern state. They are responsible for the professional quality of the services rendered. In so doing, they have the opportunity to interact with many business, labor, and citizens groups. Both policy and administrative tasks are involved, but administration is definitely more pronounced here than at the top of the political system. While there is a great deal of administrative content, it provides much room for discretion, which can be used for rendering favors as well as for combining administrative efficiency with a human touch.

The base of the political system is called the micro-level of policy making. Large number of decisions are made, but they are generally limited in application. They revolve around individuals, companies, and communities that are seeking benefits for themselves from the larger polity. Many issues can be decided by use of predetermined rules, but constant care is needed to protect the equality of all before the law. "In this kind of situation, democratic morality calls for protective arrangements that will prevent eclipse of the equalitarian principle by favoritism to selected portions of the society." (Redford, 1969: 85.) Micro-decisions can be programmed for administrative handling in different ways. Most can be handled through rules that will govern decisions of a particular type. A second approach is "to sink in administrative process the application of policy to particular situations." (Redford, 1969: 86.) Reliable records, due process, and the rights of internal review and of appeal from one administrative level to another are used to arrive at objective fact finding and decision making. A third method, reserved for the more unusual cases, "is to seek judgement by persons who will act expertly and impersonally." (Redford, 1969: 87.)

Redford's three levels of government do not correspond to the more traditional subdivisions of the political system, such as federal, state, and local governments, or the distinction between legislative, executive, and judicial responsibilities. Instead, they identify a different mix between policy making and administration, which is characteristic of different parts of the political system. This analysis leads to a more accurate picture of the complexity of political organization than the simplistic notion that decisions are made at the top and implemented by subordinate institutions. Political systems are hierarchically structured only to a limited extent — and often more so on paper than in reality. In fact, one of the weaknesses of Weber's theory of the administrative state is the assumption of a perfectly hierarchical flow of power, decisions, and information in the political system as a whole, as well as in individual agencies. Using Crozier's and Redford's findings, I shall summarize the discussion up to this point by offering some observations about the development of scholarly reflection on the development of legal rationality and the administrative state in the post-Weberian era (See also Karl, 1976).

1. Large-scale bureaucracies continue to be needed in the contemporary state, but their efficiency is no longer identical with machine-like impersonal performance.
2. Dysfunctions of bureaucratic behavior can be reformed from within to make governmental agencies more acceptable to an informed and educated citizenry. A close union between political mandate and the day-to-day discharge of agency responsibilities is of paramount importance.
3. Policy making and administration are the responsibility of all parts of the political system, though they come in different combinations and need to be dealt with in recognition of these differences.
4. Fully hierarchical administrative organization creates difficulties for citizen access and weakens staff morale. Decentralized administrative and political units are being developed in many places but do not, in their present experimental state, represent an alternative to bureaucratic organization.
5. The administrative state continues to derive its rational qualities from the systematic use of rules and procedures which are publicly known and applied to all appropriate cases.
6. While the procedural aspect of rationality continues to be heavily dependent on legal reasoning, the substance of administrative rationality has become more diversified and complex. Simon's distinction between routine decisions that are programmable, and substantive decisions that must be prepared analytically, is important (Simon, 1960). But, the nature and role of substantive information and analysis require detailed study to understand their importance.

A number of research tasks follows from this listing. Since the borderline between policy and administration is vague and shifting, analysis of the policy-administration mix in a variety of institutional settings is needed. Take the example of state governments. For a long time it seemed inevitable that their policy-making functions were eroded by the federal government, leaving the states as administrative agents responsible for implementing federally enacted and funded programs. While the trend towards nation-wide programs and standards is unlikely to be reversed, there is increasing evidence that states, if they choose to, can adopt a more active and policy-oriented stand. They can do this by using federal legislation to address, in an innovative way, the particular needs of their population. Or, they can revive the old notion of states providing a testing ground for new policies prior to national action. We found a perfect example for both options in a study of nutrition policy (Schmandt, Shorey, and Kinch, 1980). Arizona used a federal health and nutrition program to address the needs of its indigent population, among them, Indians living on reservations. Other states, like California, Connecticut, Massachusetts, and Michigan developed new ways to deal with health problems resulting from affluent diets, providing the federal government with a number of experimental approaches in its own search for a viable strategy. Similarly, European governments are searching for new ways to decentralize administrative structures and allow more active political roles to regional and local institutions and citizen groups (Nelkin, 1979 and 1980; Nelkin and Pollack, 1979).

This is part of a broader issue. As the search for new organizational structures accelerates, the promise those experiments hold for administrative reform needs to be assessed on a continuing basis. The concept of change in terms of policy making and administration will become central. The assumption is no longer valid that change is primarily restricted to the substantive areas of policy making as new laws are enacted and programs initiated. The organizations and procedures of policy making and administration are also changing, and the change in question is more than incremental (Weiss and Barton, 1980). I shall now develop this theme with regard to changes in the public sector that are a result of new circumstances associated with the increased role of science and technology.

The Scientific State: A Look at the New Agenda

As mentioned in the introduction to this paper, I use the term "scientific state" to identify functional, organizational, and procedural changes in government that are related to science and technology and have actually occurred. How science would influence the nature and tasks of the state has been the object of debate and speculation for as long as modern science has been around. Since technology became science-dependent, and thus a social phenomenon that could be shaped by purposeful planning, only a century ago, its impact on the state is a more recent topic of debate. But, the debate has been intense and ranges widely from positive to negative interpretations.

For many centuries, the role of science in society and its impact on the political system was recognized as an important issue by a small number of visionaries who had great expectations about the changes they saw coming. The resulting debate was utopian by nature. Francis Bacon, writing at the very beginning of the scientific age, contributed a lasting motto: "Knowledge is power." He was followed by the enlightenment philosophers of the French and American revolutions, for whom science was the key to human progress as well as to political freedom. The theme was further developed by the nineteenth century writers Condorcet, Saint-Simon, and Comte.

While many of the predictions of these early thinkers were utopian and speculative, they all shared an insight that eluded most of their contemporaries. Knowledge, generated through the disciplined force of scientific method, was far from being an esoteric phenomenon. It might be produced in lonely search, but it would lead to far reaching changes in society, economy, and state. Sanford Lakoff, in retracing the early debate on science and society points out how rare and revolutinary such foresight was at the time. Most famous philosophers and theorists of the last several centuries did not recognize science as a factor that needed to be taken into account. "Most of the better-known theorists assigned it no great role in society." (Lakoff, 1966: 26.) Such was the case for Hobbes, Locke, Rousseau, Burke, Adam Smith, Ricardo, Malthus, and many others.

Books and essays on science, technology, and society written during the first half of this century — such as the works by Spengler, Juenger, Veblen and Bernal, or science fiction by Orwell and Huxley — suffered from a lack of empirical content and remained essentially projections of possible futures. Some were pessimistic in their forecasts, arguing that science and technology were destabilizing forces which would bring unwanted change and enslave man and his institutions. Others were confident, seeing the liberating and enriching potential of science and technology as dominant.

However, major structural changes had already taken place. During the nineteenth century, modern laboratory-based scientific investigation had grown in methodological sophistication and institutional diversification. "In the decades between 1860 and 1920, the organization of knowledge in America was transformed, and institutional patterns were established that persist to this day. . . . American science and scholarship, once fostered and sustained by regionally isolated learned societies, became centered in an expanding network of national organizations for the advancement of specialized knowledge." (Oleson and Voss, 1979: vii.) Science-based technologies, beginning with the chemical industry, emerged during the same decades and became the prototype for the innovation-oriented growth industries of the twentieth century. The economic and political powers of a union between science and technology had been demonstrated by the beginning of the century. However, the literature on the social ramifications of science and technology did not yet reflect an understanding of the intricate web of connections that holds together scientific discoveries, technological innovations, structural changes in the economy, work force, and other social institutions, and in the political system. As long as this condition prevailed, the literature exploring the social and political impact of science and technology was to remain prescientific, anecdotal, and inconclusive. It was based more on value judgments by the authors than on analysis of real-world developments. Conclusions fell generally into two groups — asking either for more recognition and support of science so that its social benefits would be reaped, or making the case for subjecting science and technology to social controls in order to contain their destructive powers.

During the 1950s scholarly works on the impact of science and technology began to appear. Gradually, an empirical research base became available and theoretical concepts emerged that helped to advance understanding of the social consequences of science and technology. But progress was slow. Part of the difficulty in developing studies of science and society commensurate with the importance of the changes they have brought stems from a reluctance of established scholarly disciplines to turn their attention to the study of new and initially ill-defined problems. The data base for work considered to meet disciplinary standards tends to be weak, causing additional disincentives for scholars to enter uncharted land.

Under these conditions, it is not surprising that many early efforts to reflect upon the new political questions raised by developments in nuclear science, aeronautics, and electronics, among others, were written by scientists who had participated in the mobilization of science and technology during the Second World War. In analyzing their experience in advisory or policy-making roles in the institutional environments of war department, prime minister's or president's offices, Manhattan Project and other code-named crash development projects, they provided the first detailed documentation for and insight into a revolutionary new relationship between science and warfare. The writings of participant-observers in England — Tizzard, Blackett, Zuckerman and Snow — and America — Bush, Conant, and Oppenheimer — document these new developments (Crowther, 1968; Schmandt, 1966). The recollections of Presidential science advisers, such as James Killian (1977) and George Kistiakowsky (1976), though published much later, belong to the same group of publications and illuminate the new role of science in policy making.

There is a common theme to many of these writings, which is of importance in our search for the characteristics of the scientific state. They discuss the obvious topic of how new weapons were developed and what roles were played by scientists and engineers to make these developments possible. But they all agree that this was only part of the new experience. They give equal attention to procedural innovations and structural changes in decision-making and organizational arrangements. After the war, it became clear that only the Anglo-Saxon countries had pioneered these procedural changes. They were neglected in the German effort to use science for the war effort, and probably could not have been developed under the prevailing political conditions, which did not allow for sharing of planning and policy responsibilities among party, military, and scientists. Churchill drew attention to this innovative, and highly influential, use of scientists and of scientific methods in the British approach to military and political planning. "The Germans had developed a technically efficient radar system which was in some respects ahead of our own . . . It was operational efficiency rather than novelty of equipment that was the British achievement." (Churchill, 1948, chapter 9.)

The procedural changes in question ranged from training of radar crews by scientists, and the systematic use of scientists in military intelligence and in the evaluation of field operations, to their participation in strategic planning. The development of operations research as a tool in the planning of military operations belongs in this context. It all meant that scientists were used for more than strictly technical advice and became full-fledged members of the planning and decision-making teams. This created powerful new capabilities, along with serious new problems, about the appropriate roles of experts in the policy process. Using the example of scientific advice to Prime Minister Churchill during the Second World War, C. P. Snow (1961) analyzed the dangers of biased scientific inputs into the political decision-making process. He concluded that even high-quality advice must never dominate decisions, since these have to express political and moral judgements, rather than technical factors. This led Snow to his well-known warning against the usurpation of the decision-making process by a "scientific overlord." Other dangers were to emerge, such as the technology-oriented close partnership between the military and private industry. The resulting interests proved difficult to control with traditional political and administrative means and were seen as a powerful coalition pushing for ever increasing development of military hardware. This was criticized in President Eisenhower's famous farewell message on the military-industrial complex.

But, the very fact of criticism of this kind attests to the power of the new union among science, technology, and warfare. The changed nature of warfare in the nuclear age and the far-reaching impact on the nature of relations among nations have been dominant themes of the new literature exploring the policy implications of science. There was an urgent need to understand the radically changed nature of international relations and the dynamics of the arms race among the superpowers. Studies of strategic weapons, nuclear proliferation, and arms control all were addressed to this need. Scientists who had been involved in such high government positions as science adviser to the president and director of research and development in the Defense Department, wrote extensively about the need for political, rather than technology-driven decision making, in order to slow the speed of the arms race and adopt arms control measures (Wiesner and York, 1964). Given the nature of modern weapon systems and their large demands on national resources, much of the research in this field centers on relations between America and the Soviet Union. This is one reason for the heavy emphasis on American experience in the Western science policy literature. During the 1950s, the United States seemed to provide not only the dominant, but the only model for the scientific state. This has changed as other policy concerns — economic growth and productivity, health, energy, and the environment — have come to the forefront. To the extent that nuclear and strategic arms, as well as competition in space, continue to be primarily superpower concerns, the American model remains strong and offers some unique characteristics. This is particularly the case for administrative and procedural changes, such as the partnership among government, industry, and the universities; the growth of non-profit study and consulting organizations providing services to the government; and the high degree of mobility of professionals, many of them with scientific and engineering backgrounds, among different organizations in the private and public sectors.

A second major dimension of the scientific state comes into focus in the extensive literature dealing with innovation and the roles that governments have assumed in stimulating research and development through tax measures, patent legislation, and direct financing. The appropriate strategy for government intervention is quite controversial and has found different answers at different times and in different countries. Both domestic and international policy issues are involved. The United States has been reluctant to provide much direct funding for industrial innovation. European governments have taken a more active stand, often arguing that American military and space programs had important economic payoffs that needed to be compensated for by direct government subsidies to industrial research and development in such key sectors as electronics and aeronautics (Pavitt and Walker, 1976).

The impact of science-based innovation on international competitiveness has been hotly debated. During the 1960s, Europe was concerned about a widening technological gap which was explained by American leadership in space and defense as well as in business-government relations. At the time, the rising influence of innovative American-dominated multinational corporations was followed with much suspicion accompanied by a keen interest to learn from their ways of doing business. More recently, Americans have become concerned about losing scientific and technological leadership as well as administrative flexibility and easy access to risk capital (Brooks, 1972).

The prevailing policy in the United States of promoting innovation in the economy through indirect means, primarily deductability of industrial research expenditures for tax purposes, has not been followed in two areas. For more than a century, the federal government has underwritten the major part of agricultural research expenditures. This commitment began with the creation of land-grant colleges and state experiment stations, which received formula funding from the federal government. The small size of individual production units, as well as the government's early interest in modernizing the agricultural sector, dictated a different strategy compared to the industrial sector. Much of the agricultural research system remains in place to this day, often leading to complaints about inefficiencies resulting from assured formula funding and the lower quality of the scientific work performed in agricultural research institutions (National Academy of Sciences, 1973). "The fact remains, however, that . . . a system of decentralized and diversified research institutions, supported and guided by the federal government, controlled locally, and integrated into teaching institutions, proved highly successful in improving agricultural productivity by generating and applying scientific knowledge." (Schmandt, 1975: 203–204.) The agricultural model of the science-state interface demonstrated that multiple goals can be served, in addition to the immediate objective of creating useful scientific knowledge. Experiment stations and land-grant colleges played an equally important role in providing technical and general education to a large segment of the population, thus contributing importantly to an improvement of the social environment in the rural areas.

Governments, both in the United States and in other nations, have also been prepared to support innovation that is seen as useful for the development of an efficient economic infrastructure. This has been the case, for example, with regard to the development of commercial nuclear energy and of public transportation and communication systems. The rationale for governmental funding of large technological development projects is generally derived from the very size of the undertaking, making private funding difficult or impossible. This strategy requires a sound assessment of economic and social benefits that can be expected from the investment. Decisions of this kind are most difficult to make. In the case of nuclear power development, demonstration of technical feasibility has often been defined as the limit of governmental involvement, while the private sector would be responsible for the remaining stages of development. That technical and political considerations carry different weights for apparently similar decisions, according to different circumstances, is not surprising. The opposite outcomes in the American and British-French decisions on the development of a civilian supersonic aircraft illustrate this fact.

In this, as in innumerable other cases, the role of the public administrator has changed. He must use scientific and technical information, combine it with relevant economic, social, and political considerations, and arrive at a reasonable judgment about the appropriate course of public action. He continues to use legal information, as in the Weberian model, to ensure due process in reaching decisions. But he must add the new dimension of assessing, in a structured and methodologically sound way, scientific, technological, economic, and social evidence. The new tasks have changed the nature of decision making for large parts of the public sector, not only for investment decisions on new technological ventures, but for the myriad of regulatory issues that face the government in public health, environmental protection, and social policy. As science and technology, through private and public actions, change the economic and social environment, the government has become responsible for assessing the consequences and weighing costs and benefits in order that progress in one direction does not lead to unacceptable damage in another.

This first look at the new policy agenda brought about by science and technology leads to three observations. First, states have to deal with new or significantly different policy issues as a result of the widespread societal uses of science and technology. Second, the new issues do not lend themselves to neat compartmentalization, just adding to the responsibilities of governments a new rubric for science policy. Instead, the new policy issues are dispersed all over government and become an integral part of major policy functions. Third, the new policy issues tend to be complex, difficult to understand, and even more difficult to cope with in terms of finding political and administrative answers. The search for appropriate and timely responses is a major challenge for contemporary policy making and administration.

To advance understanding and handling of the new issues, it is necessary to consider both the inherent properties and effects of science and technology and the political, institutional, and economic circumstances of the various policy environments in which the issues have to be faced. Science policy cannot be studied as a separate policy function but must be placed in the context of its functional environment. This task has been only partly resolved in the scholarly science policy literature that has become available to date. The growth of scholarly activity in this area has been rapid in recent decades. A comprehensive review of the field, yielding a massive volume of over 600 pages, shows that studies of science, technology, and social change have become established on a world-wide scale (Spiegel–Roesing and de Solla Price, 1977). East and West, developed and Third World nations, large and small countries have recognized the need to analyze the new policy environment created by science and technology. But, the book also reveals some shortcomings of science policy studies. Scholars, as well as science policy officials, often view science policy as a rather narrowly defined technical function of policy making and administration. This explains a heavy focus on major scientific or technological programs, as well as on the specialized research and development budgets and institutions designed to support these activities. Typical topics of analysis include the allocation of financial and human resources to science and technology, relations of governmental, scientific, and technological organizations with industrial and academic contractors and grantees, establishment and management of national research facilities, decisions concerning controversial or dangerous scientific and technological developments, arrangements for international cooperation, and science policy advice to the head of government or to ministers in charge of major agencies. All of these are important issues in their own right, but they tend to assume a more technical and self-contained nature of science policy issues and decisions than is the case (Schmandt, 1978).

Studies of the scientific and technological components of defense, economic, energy, or health policies — to mention a few examples — have a better chance to illuminate the interplay among technical and social, economic, and political factors. Examples of this approach include the extensive literature on weapons and weapons control, economic innovation and productivity, traditional and new sources of energy, the health and environmental effects of pollution, constraints imposed on economic growth by the limited availability of nonrenewable resources. An important advance in understanding the science-policy interface will be achieved by systematically combining into one the two analytical approaches present in the literature, one emphasizing the technical characteristics of science policy as a relatively self-contained governmental responsibility, the other focusing on the changed substantive content of important policy issues. I have proposed elsewhere a research program to meet this need (Schmandt, 1977).

The research is aimed at identifying and analyzing a small number of functional science policies, each studying the science-government interaction in a major policy area. Different functional breakdowns are possible; a division in six functions seems small enough to combine related activities and large enough to identify significant differences. These six groups are:

1. National security and international relations: defense, arms control, international relations, aid to foreign nations.
2. Economic development and productivity: industry, agriculture, commerce, the labor force.
3. Development of the technical and economic infrastructure: energy, transportation, and communications.

4. Protection and development of the physical environment: pollution control, land use, natural resources, and space activities.
5. Social policy: health, education, and welfare.
6. Support and control of fundamental research and graduate education.

Science and technology play important but different roles in each of these functional groupings. The goal would be to identify, for each of the six groups, the specific role of science and technology. This would yield a detailed description and analysis of six partial science policies, each consisting of the articulated strategies and programs for the use of science and technology and their interplay with planning, program implementation, and program evaluation in this area of policy. A functional science policy can be defined as the totality of all intervention mechanisms — grants, contracts, guaranteed loans, tax code provisions, administrative rules and regulations, legislation and court decisions — aimed at promoting and regulating knowledge generation and utilization in a particular policy area, and the role of these measures as part of the overall policy agenda in the same area. Such functional science policies exist, all trying to develop a useful link, with movement back and forth, between a specialized policy function and a corresponding science policy effort. But, comparative information is not readily available. We don't know, in sufficient detail, how science policies for agriculture, health, and defense differ from each other, how they interact with their policy environments, and how well, or how poorly, the feedback loop operates that is intended to link scientific and technical with political and administrative functions. Knowledge of this kind would contribute importantly to an understanding of the scientific state at the working level of specific policy responsibilities. It would also yield insights into how different nations go about integrating science policy into the major policy functions. Comparative studies of industrial policies, nuclear power development, or environmental pollution control, have been published in recent years (Nau, 1974; Nelkin, 1977; Pavitt and Walker, 1976). But comparative works, combining issue analysis with an examination of functional science policy arrangements, are not available. It will take years before such a rich empirical documentation, based on national as well as comparative studies, will show in detail how science and technology have been integrated into the policy-making process, and how the emerging changes differ from one policy area to another, and from country to country.

The Scientific State: Structural Characteristics

While issue- and function-specific information is not as readily available as one might hope, structural and organizational characteristics of the scientific state have long been identified that make it possible to determine the general direction of change and to compare the new features to the characteristic qualities of the administrative state. Outstanding aspects of the new administrative and political system include decentralization of decision making, institutional diversification, increased interactions between the public and private sectors, and team approaches to analytical and administrative tasks. Many of these changes have been analyzed in the literature dealing with organizational theory and behavior. I shall begin by discussing authors who combine science policy and administrative experience, and, therefore, are in a better position to show connections between the new policy roles of science and their organizational and procedural implications.

Writing as early as 1954, Don K. Price, in *Government and Science,* lists several important innovations. Each of these has had long-lasting effects. There is, first, the need for scientific input into decision making. The president, and other high officials, needed direct, confidential, and readily available access to scientific expertise. The questions of war and peace in the nuclear age have always loomed large in this regard, but other issues, domestic as well as international, increasingly needed to be considered from the viewpoint of science policy. New institutional mechanisms were established to meet this need. Many approaches were tried, such as locating central science advice in the Defense Department, the National Science Foundation, or the President's executive office. Other nations created ministries for science and technology — a move often discussed in the United States but never receiving enough political support to become a realistic possibility. Beckler (1974), Golden (1980), and Katz (1978) have reviewed the American experience. They seem to agree that the president's science adviser has been influential, at times, as a source of objective technical advice, untainted by agency parochialism, and able to bring academic and industrial experts into the policy process. They also show, however, that the office has not been able to act as an early warning system that would alert the President to policy issues that are about to become critical. This was not due to a failure in identifying the right issues and subjecting them to analysis. In many instances, the reports that were prepared by the office, later proved to be unusually prescient. The failure was one of political impact and institutional clout. Harvey Brooks, who served on the science advisory board to the President, made this point: "In almost every case we failed to get the attention of top policy-makers sufficiently to raise the issue to the necessary level of political visibility to generate concern and action." (Quoted in Golden, 1980: 88.) This suggests that new institutional capabilities for science policy advice and analysis have a difficult time to be accepted by traditional government organizations. They have to be able to present technical advice in language and arguments that is accessible and convincing to non-technical policy makers. Since science advice and analysis are needed not only in the direct proximity of the chief executive, this points to the general need, throughout government, to improve the linkage betweeen technical and administrative or political staffs and decision makers.

Price also discussed the problems involved in another institutional innovation in the public sector. A set of new organizations was required to generate scientific and technical information and to manage major new technologies. In each case, the option of providing the required services internally to the government or of contracting with universities, industry, and new non-profit organizations for research, development, and analysis had to be carefuly considered. The preference in the United States has generally been to involve the private sector as a contractor for required services. Thus, one of the earliest science support agencies, the National Science Foundation, was designed as a stimulator and funding source for research in the universities. The National Institutes of Health used both internal and external research. Similarly, large technical organizations, such as the Atomic Energy Comission, the National Aeronautics and Space Administration, and later the Department of Energy, used a mixed strategy, necessitating the development of new administrative arrangements to structure the cooperative links between the government and the private sector.

The principal problem was recognized early on. Relations with universities and industry had to be flexible and should not unduly interfere with the actual conduct of scientific or technical work. At the same time, standards for the responsible expenditure of public funds had to be defined and enforced. Price illustrates the early stages of dealing with this problem in his discussion of the new uses of the government purchase contract. This simple administrative device was ingeniously adapted to the acquisition of scientific knowledge and technical know-how by the government. Thus, it became possible to contract for knowledge from universities and industry. The advantage was twofold. Scientists and engineers could remain in their normal institutional environments, which were better adapted to successful work in fundamental or applied research and development than hierarchically structured traditional bureaucracies. On the other hand, government did not have to build as large an additional bureaucracy as if it had decided to meet the new science-policy needs by exclusively following the traditional model of administrative organization. Contracting for brainpower, rather than adding more and more people to the government payroll, proved to be a revolutionary innovation. It turned out to be an excellent way for the conduct of high quality scientific and engineering work. It kept government smaller than it would have been otherwise, led to a close partnership between the public and private sectors, and exposed government to non-traditional forms of management and organization that had been developed in industries and universities.

The last point is particularly important in the present context. Both research and engineering tasks require a team approach with different skills represented on the team, a large amount of independence to develop and manage the group's program, and less reliance on hierarchical direction and control. The constant contact of public administrators with university and industry groups which used new organizational arrangements and planning tools was an important factor in the development of alternative forms of administration in the public sector. Renate Mayntz (1974), commenting on comparable changes in European public administration, speaks of "functional authority" — as opposed to Weber's legal-hierarchical authority — as the underlying principle of an administrative system dependent on expertise. To organize expertise in a policy-relevant way has been a major challenge for organizational innovation. Non-profit organizations, like the Rand Corporation and the many analysis-oriented organizations derived from this model, played a pioneering role in this respect. They first transferred the multi-skill team approach from its original home in scientific research and engineering to the public sector, and demonstrated that this form of organization was applicable to the rapidly growing analytical tasks of government (Smith, 1966; Smith, 1977).

As always, new departures give rise to unforeseen side effects. The close partnership between government and private institutions has been criticized for its diffusion of public and private responsibilities, its lack of political accountability, and its tendency to prefer costly and hardware-oriented solutions in assessing policy alternatives. The study of the so-called military-industrial complex has documented the need for not losing sight of basic differences between public and private organizations and for administrative reforms to reduce frictions and abuses in government-industry relations (Edmonds, 1975; Rosen, 1973). Similarly, government-university relations have suffered from an erosion of mutual trust and increased reporting and control requirements, many of them imposed for reasons unrelated to the immediate purpose of the cooperative arrangement (Ford Foundation, 1978; Smith and Karlesky, 1977). In public agencies, the team approach, with its horizontal communications and control network and reliance on collegial decision making, can lead to organizational inaction, delay, and confusion (Mayntz, 1974: 487).

These are serious problems which call for reforms of various kinds. But dysfunctional developments of the kinds mentioned do not seem to call into question the basic features of a more decentralized administrative system with increased public-private sector interactions, task-oriented team organization, a public service recruited from a variety of disciplinary and professional backgrounds, and a work style heavily influenced by the customs and methods of scientists. Price, in his first book, had related the early stages of these administrative changes to the government's new responsibilities in science policy. He had also shown that the new ways of organizing and working were not restricted to the immediate science-policy organizations and tasks, but tended to affect the public sector in general. This trend was discussed in Price's second book, *The Scientific Estate,* published in 1965. Price organized his analysis around three structural changes that had become more pronounced in recent years and needed to be better understood for their implications — a task for both scholars and decision makers.

First, the public and private sectors had moved even more closely toward each other. The division of labor between them was increasingly obscured. As a consequence, the problems of public accountability and control had become more pronounced. Second, the government, in many areas going beyond defense and atomic energy, was confronted with problems that were scientifically and technically complex and became even more intractable when their social and economic effects and their interactions with other equally complex problems were considered. This made it necessary for the government to acquire more substantive knowledge and to equip itself with new planning and analytical capabilities. Third, accountability, expertise, and decision making were increasingly difficult to handle with traditional administrative tools. New models of organization and decision making were needed which were capable of mastering complex substantive matters, while continuing to be responsive to older concerns about due process and administrative efficiency.

Price concluded that the three areas of change had reintroduced "a set of political problems that we thought we had disposed of forever by simple Constitutional principles." (Price, 1965: 18.) The structural and functional changes in government can no longer be dealt with by relying exclusively on legal rationality and hierarchical organization. The new task is to incorporate scientific knowledge into the process of policy making and administration. In Weber's terms, this is the question of how to combine legal and scientific rationalities. Or, in Price's words: How to relate the scientific estate to society's political purpose?

Price points out that this task is intrinsically difficult, since the principles of uninhibited search for truth and timely action to protect the public interest pursue different goals, use different methods, and require different institutions. Historical experience shows that the quality of scientific work tends to suffer when political considerations dominate. The scientific estate derives much of its social value from a critical and independent posture. The political system cannot be run scientifically. It must exercise reasonable controls over science, to account for the expenditure of public funds, and recognize and act on socially unacceptable side effects. The traditional tension between science and government can no longer be neutralized by keeping them apart from each other. Since science affects important social and economic activities, it also permeates the political system. Thus, the public sector must find ways of dealing with science and making use of it. This is a process that changes both science and government. It cannot be halted. The only question is whether arrangements can be found that do not undermine the unique strengths of each partner to the union.

"Science," writes Price, "has made executive agencies not only more independent but also more interdependent." (Price, 1965: 47–48.) Agencies look less like large bureaucratic administrations handling innumerable cases according to well-established rules. They must exercise more initiative and judgment and deal with complex and non-routine tasks, which cannot be disposed of easily, by finding the appropriate precedent and making a legal determination. The routine component of the work load has diminished and what remains can be programmed more easily for machine operations. But, substantive action requires professional and technical expertise much more varied than legal and bureaucratic skills. These continue to be required, but they must be combined with scientific, technical, and social science contributions. To develop the necessary information base for decision making requires a new work style that is somewhat less dependent on instructions received from above and relies more on structuring one's own tasks, together with like-minded colleagues, and searching for solutions to new tasks. The nature of the work to be performed, and the training, professional experience, and expectations of the staff employed do not lend themselves to traditional management styles.

What happens at the work level of the individual employee is reflected in the organization's entire make-up. Agencies are less routine-oriented, certainly at the national level, but other public sector organizations are equally affected, though often with a time lag, compared to national organizations. Management and control techniques developed in the private sector are adopted — at times without sufficient consideration of the applicability to public sector tasks, but clearly indicative of an accelerated flow of ideas and people between business and government. Agencies plan, operate, and react to new situations less like traditional bureaucracies, relying on agency-wide policies and rules, but more like problem-solving institutions, searching for appropriate solutions through a process of formulating hypotheses and conducting experiments. Decentralization becomes an acceptable way of doing business, and administrative power is no longer equated with the size of the headquarters building and the number of line staff.

These are major departures from the "old logic of administration in which a large-scale organization achieves efficiency by marshalling all technical efforts under the guidance of centralized planning." (Price, 1965: 48.) The growth of the public sector takes place, to a large extent, outside the government proper, leading to the development of program-implementation structures that consist of parts of several public and private organizations.

(Hjern and Porter, 1979). Tasks and responsibilities of the government continue to increase, in response to heightened public expectations and growing technical complexities. But growth is no longer exclusively in the "form of a vast bureaucracy, organized on Max Weber's hierarchical principles." (Price, 1965: 75.) Instead, the emerging scientific state uses new organizational principles, builds new institutions for analysis and implementation, and enters into cooperative partnerships with institutions in the private sector — all features that differ significantly from the administrative state. Organizations are often smaller, more diversified, more flexible, more knowledge-dependent, less predictable in their behavior, and more difficult to subject to political and management controls.

While changes in the direction of more decentralization, cooperative arrangements, and experimental problem solving are steadily gaining ground, applied-management sciences have introduced various management control techniques designed to rationalize decision making and keep control over goals, activities, and resources of the more diffuse organizational landscape. Planning, budgeting, and management systems of the PPB, MBO, or ZBB kind are intended to provide a counterbalance to the various decentralizing forces in an attempt to replace or supplement legal-hierarchical rationality with a new system of scientific rationality. In this approach, the reliance on scientific methods is frequently stymied by a lack of consideration for the non-quantifiable elements of policy making. Landau and Stout (1979) speak of "Type II errors" that result when surprise, risk, and uncertainty are underestimated and sought to be controlled for by reliance on sophisticated management-control techniques. Management, they remind us, operates in an environment that is unregulated, risk-bearing, and problematical. As such, it is an experimental process that searches for a body of knowledge that eventually permits control. Management techniques, by contrast, tend to assume a fully-ordered and information-rich environment. They are scientific only in a superficial sense, since they discard the need for experimentation and search for solutions. Truly scientific management recognizes critical shortage of knowledge in real-life situations. Premature programming will lead to inappropriate solutions and will undermine the morale and sense of mission of the public servant. He will thrive in an environment of mutual trust where he is part of a systematic program of experimentation, imagination, analysis, and criticism. In the same way that the administrative state, based on the principle of legal rationality, became dysfunctional through an overemphasis on legal formalism, the scientific state is endangered when it substitutes methodology and technique for decision making and management. "Time after time, control systems, imposed in the name of error prevention, result only in the elimination of search procedures, the curtailment of the freedom to analyze, and a general inability to detect and correct error." (Landau and Stout, 1979: 154.)

The significance of the changes described, promising as well as dangerous ones, can be summarized by seeing the government and its agents and institutions less as custodians of stability, and more as change agents whose function it is to formulate the implications of technological, economic, and social change in terms that are understandable to the public, and to search for appropriate public policy responses to constantly changing conditions. The task is less to administer what is known, and more, at least at the frontiers of public service, to manage the unknown. This requires a new attitude to innovation as well as a different knowledge base to understand the problems encountered. For the discharge of public policy functions in such an environment, the scientists' experience in dealing with uncertainty through a process of systematic trial and error is essential and influences organizational and work procedures in the public sector. The legally trained bureaucrat is joined by a group of people bringing different skills — scientific, economic, and organizational. The civil servant is a planner, policy analyst, coordinator, and supplier of resources. He works with concepts and information from different fields. Hierarchical organization is supplemented by horizontally organized networks for policy formulation and implementation. Agency performance is dependent on analytical skills and on frequent interaction with experts from outside the government. Integration of expertise and knowledge in the decision-making process in a politically responsible way becomes a major task (Price, 1965: 75–78).

Many of the changes in public policy and administration have been discussed in detail in the literature dealing with organizations and management. In the present context I wanted to focus on the direct connection between organizational and procedural changes in the public sector and the new social significance of science. For this task Price's contribution is important. How similar developments are discussed in the administrative literature will be documented briefly. Porter and Warner (1979) distinguish between two parallel systems of administration and management, which correspond to the Weberian traditional bureaucracy and Price's new decentralized and knowledge-oriented units. The two foci for administrative action are organizational development and service delivery. The "two orientations coexist, often uncomfortably, within public and private organizations." Hjern and Porter (1979: 30–33) deal specifically with the characteristics of service delivery agencies. Modern implementation structures combine parts of private and public organizations and they, rather than traditional hierarchical organizations, have become responsible for service delivery in many areas. "Implementation structures are viable, non-hierarchical general systems . . . comprised of an array of public and private actors . . . Traditional hierarchical authority tends to be of limited influence (for the delivery of professionalized human services). Professional competence is given deference . . . There are substructures for policy making, planning and intelligence, resource provision, intermediary and coordinating roles, service provision, and evaluating." A common characteristic of the new organizational units is their increased dependence, compared to traditional agencies, on knowledge, information, and scientific method. I shall return to this point.

The growth of implementation structures involving multiple actors, professionals from different backgrounds, and collegial work procedures, seems to be a common trend in different countries. For example, a study by Hanf, Hjern, and Porter (1977) reported the emergence of similar structures in the Federal Republic of Germany and Sweden. The new structures organize service delivery systems that are more personalized, more professional, more interactive, and often more aggressive in advocating clients' rights. They represent interesting experiments and, perhaps, alternatives to traditional service delivery by combining technically and administratively complex functions with increased flexibility, responsiveness to local circumstances, and citizens' needs. (See also, Nelkin, 1977.) These developments depart significantly from other forms of organization in the public sector. In the past, administrative theory and practice were developed around industrial, military, and other single-purpose bureaucratic organizations of the Weberian type. While "hierarchy and the analysis of single functions" were of central importance for traditional bureaucracy, the modern public sector must "be able to cope with implementation processes which are multi-organizational and discretionary." (Porter, 1980.) Many of these programs are experimental and introduced in response to field experience elsewhere and rapidly changing demands for particular services. Under these conditions, an effective administration allows decentralized and non-hierarchical units to "cope with situations where there is a necessity for local discretion, exercised within guidelines and regulations set and monitored by central management." (Porter, 1980.)

The work just reviewed comes from the fields of social services and organizational theory. The authors stress the importance of social experiments as one of the origins of flexible and client-oriented service delivery. They do not discuss, except by implication, the "scientific" aspect of the new approach to program delivery. The professional backround of many of the professionals involved in the delivery of human services — psychology, psychiatry, medicine, counseling — suggests that they have brought their scientifically oriented approach quite systematically into the delivery of publicly administered programs. In much of their work we find reliance on formal planning procedures, the formulation of hypotheses and the conduct of experiments, as well as a commitment to the value of evaluation and feedback. Other professional environments, such as those of the engineering professions, have also brought elements of the scientific culture to the public sector. This has long given a special flavor to the French administration where high civil service positions have been occupied by graduates of the science- and engineering-oriented "Grandes Ecoles." A comparable professionalization has occurred in many countries. At the present time, it is particularly pronounced in regulatory policy making where the complexities of risk to health and environment necessitate a heavy influx of biologists and chemists, among other disciplines. It seems that these changes, as well as the administrative trends discussed by Price in the context of innovations in science policy, are closely related phenomena. To discuss the common nature of these developments, I shall return one more time to Price, or more precisely, to one of his critics. This will give me the opportunity to introduce the broadened concept of the scientific state that I mentioned in the introduction to this paper.

Three Faces of the Scientific State

Boulding, in a review of Price's *Scientific Estate,* acknowledges the importance of the book for understanding significant science-related changes in governmental organization and behavior. He finds, however, that the changes at work are more fundamental and far-reaching than perceived by Price. Their real significance will come into focus, he contends, once our new social system, and the forces shaping it, are recognized as being dependent on the process of generating and managing information. From this point of departure, which is similar to the central role accorded to scientific information by Daniel Bell and others, Boulding goes on to make his principal point: Science is generally perceived to have its greatest impact in changing different aspects of society through the results of research. However, the methods of science — the systematic ways of identifying, analyzing, and solving problems — are the most powerful forces which have an enormous impact beyond the scientific arena itself. Not the end product, but the process of scientific investigation, is the really important change factor. Science shapes the contemporary world through the power of methodologically processed and ordered information. Social activities that are heavily dependent on information are most affected. Public policy and administration fall in this category. "Science derives its political importance from the fact that it has become an increasingly significant part of the whole social process by which images of the world — including the images of political decision makers — are created." (Boulding, 1966: 132.)

Boulding does not develop this observation any further in the review of Price's book. But his comment is important for an analysis of how science changes the polity. It is illustrative for the views of comtemproary authors — such as Bell, Brzezinski, McLuhan, and Simon — who see science, and in particular, its method and approach rather than its specific technological extensions, at the root of major changes in society and government. Others disagree with this view. Most science policy literature, for example, sees a more limited impact of science on the structures and processes of government. After a careful review of that literature, Lakoff concludes that "it is difficult to know with certainty whether the problems posed by advances in science and technology will compel significant and long-lasting changes in domestic and international political systems, or whether any changes will prove to be temporary." (Lakoff, 1977: 385.) However, Boulding's focus on scientific method suggests that the change potential of science goes beyond new research results and technology, or, as Lakoff puts it, "advances in science and technology." By viewing science as a multi-faceted change agent, structural and procedural changes in the political system come into focus. It is thus possible to conclude that the system has already undergone major changes, as our analysis contrasting administrative and scientific states tried to demonstrate. There are three aspects of this change process. Each reflects a different power or quality of science. I distinguish these three modes of science by their principal functions in policy making and administration: science as product, science as evidence, and science as method.

Science as product is the power of science to lead to new things, or ways for doing things, that are seen as socially desirable. It is this function of science that is most prominent in discussions of the science policy responsibilities of the modern state: Since science enlarges the options available to the decision maker by making it possible to do things that were physically impossible to do previously, public support for scientific research is provided. The support goes to science as a production force. Many of the end products of science are not socially useable until they are embodied in technologies and their attendant social and economic systems designed to produce, market, use, and control the new technologies. It is distinctive of our age that this process of innovation depends on science, and science policy came into being as this process was understood and the role of government in supplying resources and controlling the uses made of them was accepted. The products made possible by science, and calling for government intervention in the production process, cover the range of activities listed in our earlier discussion of functional science policies — weapon systems, new sources of energy, hybrid seeds or lean pork, machines for kidney analysis, drugs for the treatment of hypertension, and so forth. (For overview see Table 2.1)

Other products of science become socially useable without being transformed first into technological products or processes. Such is the case of science-based practical knowledge, such as medical know-how and computer programs, management information systems, and research data. The distinction between these two classes of scientific products — those making use of science embodied in technology and those making practical use of science-generated information — is important, since it calls for different policy responses. Financing and managing the production process, as well as controlling social and economic impacts, are different for each group. But both classes of products share the fundamental quality of being science generated and science dependent. It is for this reason that science policy is appropriately named, even though technological developments are often more visible than the science behind them. Science as product has resulted in massive public sector initiatives, leading to policies and programs designed to derive public sector benefits from science, or to control dangers and disbenefits of scientific products.

Policies to meet these objectives, and public sector organizations and procedures serving them, have developed on a world-wide scale. But, larger countries tend to be more heavily engaged than smaller nations, and industrialized countries more so than less-developed ones (Long and Wright, 1975). This is parimarily due to the fact that large and advanced countries are most heavily involved in military and space research and development. They are also more likely to devote large resources to high technology programs in such areas as energy, transportation, and communications. The availability of competent academic and industrial sectors as participating partners in developing national science policies provides an additional edge to the larger industrialized nations. For a long time, encompassing the entire period between the end of the Second World War and the late 1960s, the United States was seen as the universal model for developing a national science policy effort built around the productive function of science. The model was unquestioned in the Western countries but proved equally intriguing, for its apparent success, to the Soviet Union and its allies. The United States objective was "to achieve rapid scientific advance and to exploit the advance for social, economic, and military purposes." (Gilpin, 1968: 125.) Gilpin, in analyzing the principal features of the American approach, identifies a number of "imperatives" of the "modern scientific enterprise and the new role of science in technological advance." A successful national science policy must recognize these imperatives.

One of these imperatives, according to Gilpin, is the need to bring the conduct of research from the small-scale laboratory environment of the nineteenth century into the industrial age. Research must be industrialized — brought to a larger scale, use teams of researchers with different skills, and have large resources at its disposal. This is a process not limited to the organization of research in industry, even though it was pioneered in this environment. The same principles apply in the case of universities and government where large, multi-disciplinary, and well-managed programs developed rapidly during the post-war period. Such a policy of growth was based on a recognition of science as a national resource and an understanding of the specific organizational requirements conducive to high quality scientific work.

Gilpin describes the second characteristic of the American model as the "nationalization of research and development." In part, this refers to the dominant role of government in funding research and development for defense and space purposes. But it also refers to the emergence of new institutional ties between the public and private sectors, which I discussed earlier. As a result, government influences the pace of new knowledge, in particular in those areas that are judged to be of highest social utility. Science, though primarily conducted in private institutions, is operated as a decentralized national enterprise. At the time Gilpin wrote (1968), the federal government funded three fourths of all research and development activities in the United States. Since then, the government's part has declined to about half, with industry financing the other half. This reflects two trends: The government has scaled down its space program, and has been conservative in launching new technological ventures of comparable size. At the same time, industry has continued to increase its investment in research and development at a regular pace, while the government, during much of the 1970s, reduced growth rates. But even so, the government's influence remains large.

Other aspects of the American science policy model include the development of a "somewhat diffuse but nevertheless identifiable mechanism for the governance of science" (Gilpin, 1968: 134), the growth of service-oriented universities capable and willing to respond to new demands, an easy flow of experts among the institutions of the scientific system, whether located in government, industry, or university. Other nations were concerned that the American economy derived large benefits from government support of military, space, nuclear, and electronic technologies. They often tended to overrate the economic impact of military products for civilian markets as a result of these programs, as was the case during the heated debate about the technological gap separating Europe from the United States during the late 1960s. But they had a valid point in emphasizing the more indirect effects on production techniques, quality control, and management throughout the engineering industries. "The research revolution which began in the national security and space industries has generated a new dynamism and set of attitudes throughout the rest of the economy." (Gilpin, 1968: 142.) American corporate management was increasingly entrusted to a skilled technical elite. An organizational and managerial revolution took place in many sectors of the economy. This was the result of new organizational knowledge and of applying scientific methods to problems of decision making and management — the domain of science as method, which I shall deal with last. But Gilpin's reference to these developments in the context of discussing science policy serves as a useful reminder that the three aspects of science-related changes are closely interrelated.

In recent years, other industrialized nations have caught up with the American advance and have developed their own science policy structures

TABLE 2.1   The Scientific State I: Science as Product

Rationale	Science enlarges the range of policy options. It must be supported and controlled as a national resource.
Policy approach	Functional science policies as an integral part of governmental responsibilities for: Defense and international relations Economic policy (industry, agriculture) Economic infrastructure (energy, transportation, communications) Physical environment and resources Social policy (health, education, welfare) Fundamental research and graduate education
Intervention mechanisms	Tax code provisions for research and development investments Patent and licensing arrangements Government funding of research and development Guaranteed loans and risk capital Control of dangerous scientific and technical activities
Organizations	Scientific agencies (support and conduct of research) Technological agencies (nuclear energy, space) Mission agencies with scientific and technological programs (defense, energy, transportation, health, etc.) Central science policy direction and coordination
Procedures	Heavy involvement of industry, universities, and non-profit organizations through use of purchase contracts and grants Emphasis on project funding instead of institutional support (United States) Parallel funding systems for projects and institutions (Europe)
Reach	Advanced nations; developing countries depend on technology transfer Heavy emphasis on defense and high-technology in largest nations Increased emphasis on economic, energy, and health concerns in many nations Requires fully developed research system

and programs. This process was particularly pronounced during the 1970s. The American lead has become significantly smaller and in some sectors it has disappeared entirely. The situation of the late 1960s has been reversed, as concern has been voiced in the United States about the economic and military consequences of a possible loss of scientific and technological leadership in the world. At the beginning of the decade, when America's investment in science and technology was declining and Europe was still increasing its commitment, Brooks observed: "I believe that the United States is experiencing only a few years earlier some of the forces and trends that will become worldwide among industrialized nations: saturation of the population able to undertake science and technology, competition of social welfare and other public expenditures for the government budget, increased public preoccupation with the side effects of technology." (Brooks, 1972: 118.) The experience of the last years has borne out this assessment. The major industrial nations seem to have reached similar plateaus, both in terms of the intensity of the research effort and of their still incomplete ability to fine-tune the link between scientific effort and economic and social outcomes. Attempts to find satisfactory answers as to where governments should intervene in industrial innovation "all run into the following difficulties: dealing with multiple policy objectives; assessing national costs and benefits; comparing with alternative policies; choosing appropriate policy instruments." (Pavitt and Walker, 1976: 14.) With the exception of large remaining differences in the military field, science policies of industrialized nations tend to look somewhat more alike, all struggling with complex and often new problems. At one point, there seemed to be rather clear-cut differences that determined national strategies for science and technology, such as exploration of all promising leads, specialization in some key sectors, importation of know-how. (Gilpin, 1970: 442). But these strategies have become more diffused and less certain, as even the largest nations recognize the restraints imposed by resource limitations and smaller and medium-sized countries can, over time, build a more balanced national research and development system (OECD, 1972–1974). A recent report by the Organization for Economic Cooperation and Development comes to the sobering conclusion that after decades of searching for workable policies, we are still at the beginning: Innovation policies have been successful only in come sectors, for example in agriculture and fundamental research. In other cases, they have been ineffective or unduly expensive, such as in subsidizing technological ventures without reference to market considerations or society's needs. "It is imperative . . . to devise new innovation policies for the present and future. . . . Research and innovation policies must be better integrated with other aspects of governmental policy, particularly with economic and social ones." (OECD, 1980: 15.) This points to the need, referred to earlier in the discussion of functional science policies, to deal with science policy issues less as self-contained policy questions, but more as integral parts of the major policy areas. This need, so far, has been considered in the context of science as product. It will now be discussed in regard to the role of science as evidence in the development of public policies.

Science as evidence refers to the increasing need in policy making to identify and analyze social problems with the help of information provided by empirical research. Whether asbestos in the work place or the general environment causes lung cancer, or the diet of affluent nations rich in saturated fats and energy contributes to cardiovascular diseases, or new industrial chemicals have toxicological effects on man and nature — questions of this kind can only be answered after extensive and time-consuming research has been conducted. What is more, only through research of this kind can the problem be recognized in the first place. For a long time, science has been used for the purposes of problem identification and analysis, and more than a century ago it led to public policy responses based on such evidence. One need only think of public health measures that were taken to improve sanitation and water supply, once the role of harmful bacteria in spreading infectious diseases had been discovered. Other well-known examples are control measures, taken early during this century, to prevent food poisoning or to combat tuberculosis as one of the major killer diseases. More recently, the discovery of vitamins and trace minerals as essential components of the human diet led to significant advances in food and health policies. The number of issues requiring scientific investigation prior to a meaningful policy intervention has increased dramatically over the course of the last two decades. The issues encountered are more complex, and effects tend to be long-term and difficult to detect. What used to be a rare need for policy makers and administrators, has become the rule in many areas of health, energy, environmental, and social policies: Without a solid base of empirical research findings, policy making is as blind and helpless to find solutions as were attempts to conquer the plague during the Middle Ages. (For an overview see Table 2.2)

It is only in response to this increased dependence on science for policy that a systematic effort has begun to understand and improve the process of preparing and utilizing scientific evidence for the purpose of informed policy making. The process of generating scientific evidence for policy-making purposes is complex and as much in an experimental stage as most of the policies devised to make use of science as a productive force. It is equally unclear, and often controversial, where and how the government should intervene, once scientific evidence is available. Should cigarettes be prohibited, or the cigarette tax be increased, or tobacco subsidies to growers be repealed? (Friedman, 1975.) These are political issues — they cannot be resolved by scientists, but they are likely to be answered differently, depending on the nature and quality of scientific evidence about the health hazards of cigarette smoking. The same goes for the future of nuclear-power generation, the role of coal in the nation's energy strategy, the admissibility of artificial sweeteners, and a myriad of similar questions: In each case, the policy response will be influenced, though not determined, by the best available knowledge about risks associated with nuclear energy, the effects of coal combustion on climate, weather, and health, or the dangers of saccharine for the general population and for diabetic or overweight persons. Scientific evidence cannot provide the answer about the correct course of action that

TABLE 2.2   The Scientific State II: Science as Evidence

Rationale	Complex policy problems cannot be solved without data base derived from scientific evidence
Policy areas	Primarily health, environment, occupational safety, food and nutrition, consumer protection. Also, some areas of social policy, such as education, vocational training, and welfare.
Methods	Surveys, epidemiological studies, scientific tests Assessments of available scientific information and its quality Assessments of health and environmental effects Social experiments and demonstrations Forecasting techniques
Intervention mechanisms	Regulation of hazards in the environment and workplace Regulation of classes of products, such as industrial chemicals, pesticides, drugs, food additives, consumer products Public information Prevention strategies Development of non-hazardous alternatives
Organizations	Science and health assessment units in executive agencies Regulatory agencies concerned with health, safety, and environmental matters Science and health assessment organizations, such as Office of Technology Assessment (U.S. Congress), National Academy of Sciences-National Research Council, public interest groups, international organizations, professional scientific organizations Non-profit and commercial contractors
Issues	Quality control of science and health assessments Decision making under conditions of uncertainty Controversy over evidence, either because evidence is inconclusive, or facts and interpretation are merged Premature or ill-advised action on the basis of partial or preliminary evidence
Reach	All nations, but with different emphasis depending on stage of development and research system Small advanced nations can participate fully, even take the lead in specific areas. Developing countries affected primarily for public health and agricultural development.

needs to be taken, but it is a necessary condition, which provides guidance to the decision maker. It alone can provide information necessary to understand the problem, assess its significance, and estimate its social and economic costs. At the same time, the processes of providing scientific evidence and interpreting its meaning are beset with problems.

How does science become an evidential base for decision making? The starting point, obviously, is scientific research in the form of experimental studies on, say, the nature, properties, and effects of toxic substances. Large numbers of individual studies are conducted, all dealing with a particular aspect of an issue. The results of these studies are not immediately useable for policy purposes. As a first step, currently available knowledge on a given topic needs to be aggregated. This is the task of a science assessment. In its simplest form, it consists of a careful review, based on literature search, of research performed and published over a period of time. A state-of-the-art survey summarizes a body of new data and structures the information by identifying different research strategies and the emerging clusters of findings: A puzzle has to be transformed into an intelligent picture. The synthesis of findings then moves on to ask questions about the quality of the new knowledge: Are results confirmed, preliminary, or controversial? What additional research is needed to fill gaps in knowledge or to eliminate bottlenecks standing in the way of future progress in research?

Scientific synthesis, in order to be useful, cannot be done as a mechanical task. Experience and judgment are required. Senior experts, or groups of experts, are best prepared to bring order to a large body of information. Other qualities required are objectivity, clarity, detachment, and the ability to present results in such a way that they can be understood beyond the immediate peer group. This last quality is of particular importance when the purpose of the assessment is not to inform the scientific community about the state of research in a given area, but to provide guidance to nonspecialists whose task is to use the scientific data base for the development of new policies.

Science assessments for policy purposes are needed in large numbers. They are often prepared under contract for agencies of the government or Congressional committees or offices, such as the Office of Technology Assessment. Professional scientific associations provide one of their most important public services in this area. The most authoritative assessments, perhaps, are prepared by groups assembled by the National Academy of Sciences. The Academy has given considerable attention to the requirements for providing unbiased and high-quality assessments. Even so, published reports are, at times, criticized for being one-sided. This points to a general problem: While science, over the course of many centuries, has developed an efficient quality control system for the validation of research results, comparable quality standards for science assessments are still preliminary. Other countries experience similar problems, all searching for different ways to improve the quality of the assessment process and for separating the synthesis of strictly scientific information from interpretation of social effects. Institutionally, the problems of quality control seem to be most severe in two instances: Commercial contractors tend to rely heavily on computerized data banks for compiling science assessments, often failing in the effort to aggregate diffuse information into a coherent picture and to exercise informed judgment about the quality of research results. Secondly, assessments prepared by in-house staff of government agencies are of widely varying quality, are often not subjected to review by outside scientists, and may, on occasion, be built around partisan political concerns.

The science assessment task, in addition to synthesis and judgment on the quality of research findings, has a third dimension. This is the assessment of health and environmental effects on hazardous substances. As long as the principal concern was with acute toxicity of pollutants, the tasks involved were straight-forward. With incereased knowledge about long-term dangers resulting from exposures to low levels of toxic substances in the work place, in the media surrounding us, or in industrial products, the accurate assessment of health effects has become extremely difficult. This is the case, for example, in assessing cancer risks, 60 to 90 percent of which have been estimated to be caused by exposure to chemical substances and, to a lesser extent, radiation (Doniger, 1978: 10). Our knowledge is limited in several respects. Not enough is known about how chemical carcinogens operate. Available research techniques do not allow for exact prediction of human risks from exposure to carcinogens. "Neither experimental nor theoretcial analysis can give . . . precise estimates of the risks associated with low doses of substances that are known to cause cancer in humans or animals at higher doses. . . . At present, the best available techniques produce only broad estimates of the outer limits of risk." (Doniger, 1978: 15.) Coping with uncertainty, without the luxury to postpone action until these have been resolved, is the rule of the game. This is even more the case in the third mode of science-government interface, since it is directly linked to the action phase of decision making.

Science as method encompasses the various analytical activities designed to prepare or evaluate public policy decisions. Policy analysis·deals with more than scientific information. It provides the linkage between scientific evidence and economic, legal, and social considerations which have to be taken into account in identifying and assessing the costs and benefits of action alternatives. Policy analysis has three main characteristics: It uses scientific concepts and methods. Its purpose is to facilitate decision making, not to generate knowledge. It is a tool for decision making, but not a substitute for it. Given these three characteristics, policy analysis occupies an intermediary role between the realms of science and action. It makes use of scientific concepts, theories, paradigms, methods, as well as empirical research results. Historically, a number of approaches can be distinguished. (For an overview, see Table 2.3)

Studies of public sector organizations and functions came into being, early in the century, in connection with the new administrative sciences. The assumption was that administrative action could be subjected to analysis of

TABLE 2.3   The Scientific State III: Science as Method

Rationale	Decisions in public policy need to be prepared and evaluated with the aid of policy analysis
Goals	Define problem, introduce empirical data base, identify alternative solutions, assess their advantages and disadvantages, recommend course of action
Approaches	Administrative efficiency and financial accountability: study of organizational structures and functions Programmatic policy planning Preparation and evaluation of programs and decisions Problem identification
Scientific inputs	Administrative and managerial sciences: organizational theory, management concepts, cost accounting Social sciences: policy concepts, substantive research results, program evaluation Mathematics, physics, economics: decision theory, quantitative analysis, operations research, systems analysis, cost-benefit analysis
Products	Decision analyses, impact statements, technology assessments, risk assessments, budget planning and analysis, program evaluations, planning documents
Organizations	Planning units in executive departments Congressional agencies (General Accounting Office, Congressional Budget Office, Office of Technology Assessment). So far, mostly in the United States Central planning staffs (Office of Science and Technology Policy, Council of Economic Advisers, Council for Environmental Quality, Office of Management and Budget). Comparable units elsewhere Independent organizations for policy analysis (organizations derived from the Rand model)
Issues	Quality of analysis. Substitution of analysis for judgment Linkage between normative and analytical statements Impact on decision making
Reach	All nations, depending on training of civil servants, interaction with policy analysts outside the government, acceptance of changes in bureaucratic work styles, recognition of new tasks and approaches to solve them

its component parts and could provide the basis for improving public agencies and their work. The New York Bureau of Municipal Research pioneered this approach, later followed by the Maxwell School of Citizenship and Public Affairs. Initially, public administration at the local level was studied. Analysis of state and federal governments was added later, leading, among other things, to the development of a presidential budget covering expenditures of all executive agencies. Social science survey techniques were used to scrutinize specific government programs. The strength of the organizational study of government was the detached analysis of agency structures and functions; its weakness was the separation of administrative from political functions, since only the former were thought to lend themselves to objective analysis.

Programmatic policy planning, though generally associated with socialist and communist practice, has also flourished in other political systems. An early example is the work of Franklin D. Roosevelt's brains trust, composed of social scientists, during his first presidential campaign. The group translated the candidate's general political ideas into a coherent New Deal philosophy and specific programs for its implementation. The detailed knowledge of the group members in different policy areas shaped the substantive content of the new programs. A similar role was played by social scientists in preparing the anti-poverty programs of the Kennedy and Johnson administrations. In these, as in other examples, an informal network of close cooperation between scientists and government officials resulted in significant policy developments.

Mathematicians, physicists, and economists have developed formal decision aids in the form of operations research, systems analysis, cost-benefit analysis, and risk assessment. These approaches are based on quantitative analysis and use of models. The methods used are scientific, but the information base is incomplete, since urgent decision needs are to be served. Quantitative analysis has been put to many uses. An early example is the Flood Control Act of 1936, which mandated the use of cost-benefit analysis as a step in reaching decisions on federal funding of new water development projects. During the Second World War, operations research was developed in the United Kingdom to plan and evaluate military operations. After the war, analysis became an indispensable part of strategic planning for the development, deployment, and use of modern weapons systems.

Another form of analytical contribution to the policy process is the identification and structuring of new policy issues. The ramifications of the automation problem for employment, training, and welfare policies were not understood until the Presidential Commission on Technology and the American Economy identified the underlying structural causes of the problem. In a similar way, the scope and seriousness of environmental dangers from widespread use of modern pesticides was not recognized until Rachel Carson wrote *Silent Spring*. In a classic chain of events, this triggered enormous media attention, public concern, and governmental action. Three years after the book was published, the President's Science Advisory Committee published a report that analyzed the problem and charted the course of required public action (Graham, 1970; PSAC, 1965). Within a few more years, federal legislation had been passed that made development and sale of pesticides dependent on meeting safety and health standards (National Academy of Sciences, 1980). In welfare policy, the development of an objective yardstick for identifying the poor and equitably assessing their needs was a pre-condition for a state- and nationwide policy response. The concept of property line, first developed toward the end of the last century by Booth and Marshall, and redefined in light of the modern social conditions in America by Orshansky, provided such an orientation aid for policy initiatives. British welfare policy, as well as the American Great Society programs of the 1960s, depended on this tool for bringing a certain degree of objectivity to a sensitive policy question (Leiby, 1978; U.S. Department of Health, Education, and Welfare, 1976). All Western nations significantly altered their fiscal policies after Keynes had developed the concept of deficit spending. The list of examples is endless. The principle is important: New ideas, concepts, and scientific tools influence policy development. It is rare, however, that this process can be foreseen with any accuracy, the delays involved be estimated, or the exact pathways from ideal to action be retraced.

The preceding discussion is based on a study of the different functional uses of policy analysis and their historical origins (Schmandt, 1981). The study shows that policy analysis has developed in different institutional environments, serves a variety of policy functions, and makes use of the methods and concepts of different scientific disciplines. There are different types of policy analysis. Functionally, for example, they serve the needs of administrative and budgetary analysis to improve efficiency and accountability, programmatic analysis to advance new policies, decision analysis to rationalize the decision-making process, or conceptual analysis to provide an intellectual handle for dealing with complex issues. Prospective as well as retrospective studies are involved, the former helping the policy maker to deal with his constantly changing tasks, the latter providing feedback from past experience for new decisions. Organized efforts to develop science-based, action-oriented tools for administration and policy began early in this century. Growth has been rapid over the course of the last two or three decades. Today, it is performed in virtually all units of government and at all levels of government. It has also led to the creation of numerous commercial and non-profit policy-oriented study organizations (Meltsner, 1976; Public Administration Review, 1977; Rein and White, 1977; Weiss, 1977).

As yet, not much is known about the quality of the work that is produced, in particular since much of it is prepared in agencies of consulting firms, does not get published, and is not formally reviewed by other experts. Not much is known, either, about the impact of policy analysis on the quality of decisions. There is an inverse relationship at work: The closer the analysis is performed to the decision point, the more difficult it gets to assess its quality and impact. Analyses conducted in more traditional institutional environments, such as universities and independent research organizations, are more easily measured for their scientific quality, but are equally difficult to weigh for their decision impact. To apply scientific quality control to agency policy analysis may be impractical, since it would lead to delays, thus reducing the value of the analysis to the decision maker. By its very nature, much of the analytical work in agencies, which should be most responsive to immediate decision needs, is prepared under time pressure, on the basis of incomplete information, and with limited involvement of the public and of independent experts. Secrecy is often invoked and, at times, needed.

Other control mechanisms in addition to scientific peer review create similar problems or are not adequate to the substantive nature of the task. Such seems to be the case of judicial review of administrative decisions or of auditing procedures to control public expenditures. Policy analysis is not a mechanical task, nor can it be performed according to standard rules and procedures. It is based on normative judgments and relates preferred solutions to preferred values. Only the process linking value assumptions to action recommendations represents the scientific part of the analysis. It is this part that can be subjected to a formal process of quality control, which must be structured in such a way that the action focus of policy analysis is not unduly impaired.

Given this constraint, the emphasis on improving quality of agency-prepared analytical work should be on the development and observance of "good analysis practices." At times, this might be supplemented with retrospective evaluations of particular analytical studies. The development of professional standards would try to streamline the process of analysis by identifying a number of specific requirements, such as the need for spelling out assumptions made in structuring and delineating the analysis, using statistical and other data according to accepted procedures, linking the conclusions reached to the body of evidence presented, and striving for logical consistency throughout the work. The quality of policy analysis depends on how satisfactorily its two constituent components are linked to each other — the judgments about desirable outcomes and ways to get there, and the presentation of the evidential data base derived from scientific, social, and economic research. This cannot be done with scientific precision; it requires judgment, intellectual honesty, and an understanding of the practical world. There is as much science as art in the process. Responsible and useful analysis will not attempt to circumvent the difficulties of such a merger by becoming either entirely scientific or judgmental. The goal of good analysis, and the standard by which it needs to be judged, must be responsiveness to the needs of public servants who must take action on difficult issues; they need all the help they can find from scientific knowledge and methods without, however, using these tools as more than an information base and a decision aid. The responsibility of the public servant, today as in the

Weberian state, is to search for just and workable solutions to problems in public policy and administration. To discharge this task under contemporary conditions requires different skills and procedures, among them the careful use of science as method.

Conclusion

Consideration of three distinctive roles of science in government — as product concerned with useful innovation and its consequences; as evidence in order to base public policy decisions on reliable information; as method in trying to develop new forms of rational decision making — has brought into focus a range of new activities, organizations, and procedures involved in policy making and administration. Each mode of interaction represents a specific set of changes and a different mediation between the worlds of scientific knowledge and public action. I have tried to show that technology and practical know-how, based on scientific information, play this role in the first case. The government gets involved as a direct user of these powerful forces and, under the general welfare clause of the constitution, in order to protect the public from possible harm. In the second case, synthesis and interpretation of scientific research results are required to understand new policy issues and devise policy responses. In the last instance, the governmental processes involved in preparing and implementing decisions make use of scientific methods and techniques to clarify the steps involved and to explain and justify choices and outcomes.

Important changes have been brought to the public sector — new, different, and more complex tasks are addressed; specialized organizations have been created and existing ones have changed; relations with the private sector have been affected, resulting in a new division of labor between public and private institutions; training, skills, and work styles of civil servants are different; and the procedures used by government to discharge its responsibilities have changed in many ways. Table 2.4 summarizes the major changes that distinguish Weber's administrative state from today's scientific state.

The changes from one state to the next are major, but they did not come at once. They have taken time, and many are still incomplete or experimental. Thus, the two models should not be considered as opposing forms of organizing for the governance of society. They are, rather, evolutionary steps in the development of rational forms of statecraft. Early in this paper, I introduced the concept of the scientific state as the attempt to merge legal with scientific rationality. As a result, legal rationality is still important, but is no longer the only road toward rational administration and policy making. This loss is also a gain: The administrative state lost some of its rigidity,

TABLE 2.4   Characteristics of the Administrative and Scientific State

	Administrative State	Scientific State
Basic principles	Legal rationality, equality before the law, due process	Scientific rationality, functional expertise, informed decision making
Methodology	Search for legal principles or precedents, adjudication	Use of scientific evidence, analysis, problem-solving techniques
Tasks	Service needs of industrialized and urbanized society	Service needs of society relying on science-based technological innovation and coping with its side effects
Organization	Large hierarchical bureaucracy, limited diversification	Decentralized functional networks, public-private partnership, team approach
Skills	Legal reasoning, impersonal and detached administration	Multiple professional backgrounds, able to cope with change and uncertainty
Knowledge base	Legal and administrative sciences, public health and engineering (limited)	Many scientific disciplines, primarily mathematics, natural sciences, economics
Pathology	Legal formalism, bureaucratic absolutism, resistance to change	Over-reliance on experts and analytical techniques. Shoddy analysis
Advantage	Capable of dealing with large volume of decisions in equitable way	Capable of identifying problems and possible solutions. More comfortable with change
Reach	All political systems intent on modernization and industrialization. Requires nationwide and centralized legal and administrative systems	All political systems. World powers most heavily affected because of imperatives of scientific warfare. Economic and social aspects dominant for other nations. Closely related to stage of economic and social development

which resulted from exclusive reliance on legal rationality, even when the increased technical complexity of modern life was no longer adequately dealt with by this principle. Stinchcombe (1959–1960) once suggested that Weber was mistaken in identifying rational administration with bureaucratic administration. He argued that there were different kinds of rational administration, with bureaucracy being a subtype; other forms of professional training and standards were capable to achieve similar results. Stinchcombe developed this thought in the context of comparing the administrative practices of mass production industries with those of the construction industry. But his insight is valid in the present context and suggests that supplementary forms of rationality — legal, scientific, and professional practices of the medical and counseling professions — have applicability to the tasks encountered in the public sector. This is a topic to be developed elsewhere. Suffice it to say here that legal and scientific rationalities, each in their own task environments, can reinforce each other. As a result, the new political system is no longer entirely dependent on rationality derived from legal analysis and introduced hierarchically through central planning and discharge of administrative tasks.

The discussion also suggests that the scientific state, as it takes hold in different nations, is less likely to be dominated by the concerns and approaches of the major world powers. They dominated, in the decades following the Second World War, because they alone had the resources necessary to exploit the new military and high-technology options made possible by scientific advances. Resource constraints, while by no means trivial, seem less forbidding for making use of science as evidence and method. This would be the case, in particular, for medium-sized and smaller advanced nations, since they have the scientific infrastructure needed, as well as a competent and diversified civil service. There are indications that some new policy concerns, relating to public health and the state of the environment, have found the necessary combination of political will and technical competence in Scandinavian countries at an earlier point in time, and with more promising results, than elsewhere. The development of modern dietary guidelines and preventive health strategies are cases in point. But without results from comparative studies explicitly looking for answers to these questions, it is only possible, at the present time, to mention the potential for different nations to build somewhat different roads leading to the scientific state. They will use similar approaches, but set priorities in light of their own conditions and concerns. On the other hand, since nations all over the world tend to encounter similar problems, many of which are directly related to man's interventions in his physical and social environment, it seems highly likely that the characteristics of the scientific state will emerge on a world-wide scale. National differences will be expressed in the specifics of organizational and other arrangements, but the principal features will be the same.

In writing this paper, I have often thought of the reader who, instinctively, will object to the term "scientific state," since it suggests loss of human control. But there is a difference, I think, between a political system being dominated by science and one having to deal with it, since it is such a significant part of the contemporary world. The modern state, in ways that Weber could not foresee, is shaped by this experience.

Acknowledgments

I thank the following individuals for helpful comments: David Porter, Lodis Rhodes, Joseph S. Szyliowicz, and David Warner.

References

Barker, Ernest. 1966. *The Development of Public Services in Western Europe*. Hamden, Conn.: Archon Books. (1st ed. 1944, Oxford University Press)

Beckler, David. 1974. "The Precarious Life of Science in the White House." *Daedalus* 103 (Summer 1974): 115–134.

Bell, Daniel. 1973. *The Coming of Post Industrial Society. A Venture in Social Forecasting*. New York: Basic Books.

_____ . 1976. *The Cultural Contradictions of Capitalism*. New York: Basic Books.

Bendix, Reinhard. 1962. *Max Weber: An Intellectual Portrait*. Garden City, NY: Doubleday and Company, Anchor Books.

Boulding, Kenneth E. 1966. Review of Don K. Price, "The Scientific Estate," *Scientific American* (April 1966): 132

Brooks, Harvey. 1972. "What's Happening to the U.S. Lead in Technology?" *Harvard Business Review* 50: 110–118

Brunner, Otto. 1943. *Land und Herrschaft. Grundfragen der territorialen Verfassungsgeschichte Oesterreichs im Mittelalter*. 3rd ed., Bruenn-Munich-Vienna. (4th ed., 1959)

Chapman, Brian. 1959. *The Profession of Government: The Public Service in Europe*. London: Unwin University Books, 1966. 2nd edition.

Churchill, Winston, S. 1948. *The Second World War. Volume 1. The Gathering Storm*. New York: Houghton Mifflin Co.

Crowther, J.G. 1968. *Science in Modern Society*. New York: Schocken Books.

Crozier, Michel. 1954. *The Bureaucratic Phenomenon*. Chicago: University of Chicago Press.

Doniger, David D. 1978. *The Law and Policy of Toxic Substances Control: A Case Study of Vinyl Chloride*. Baltimore: Johns Hopkins University Press. Baltimore and London.

Edmonds, M. 1975. "Accountability and the Military-Industrial Complex." In Bruce L. R. Smith, ed., *The New Political Economy: The Public Use of the Private Sector*. New York: John Wiley and Sons.

Ford Foundation. 1978. *Research Universities and the National Interest: A Report from Fifteen University Presidents.* New York: Ford Foundation.

Friedman, Kenneth M. 1975. *Public Policy and the Smoking-Health Controversy.* Lexington, Mass.: Lexington Books, D. C. Heath.

Friedrich, C. J. 1939. "The Continental Tradition of Training Administrators in Law and Jurisprudence." *The Journal of Modern History* 11: 129–148.

Gilpin, Robert. 1968. *France in the Age of the Scientific State.* Princeton, NJ: Princeton University Press.

_____. 1970. "Technological Strategies and National Purpose." *Science* (July 31, 1970): 441–448.

Golden, William T., ed. 1980. *Science Advice to the President.* New York: Pergamon Press.

Graham, Frank, Jr. 1970. *Since Silent Spring.* New York: Houghton Mifflin Co.

Hanf, Kenneth; Hjern, Benny; and Porter, David O. 1977. "Networks of Implementation and Administration for Manpower Policies at the Local Level in the Federal Republic of Germany and Sweden." Discussion paper 77-16. Berlin: International Institute of Management.

Henry, Nicholas. 1975. "Bureaucracy, Technology, and Knowledge Management." *Public Administration Review* 35: 572–578

Hjern, Benny, and Porter, David O. 1979. "Implementation Structures: A New Unit of Administrative Analysis." A discussion paper. Berlin: International Institute of Management.

Karl, Barry D. 1976. "Public Administration and American History: A Century of Professionalism." *Public Administration Review* 36: 489–503

Katz, James E. 1978. *Presidential Politics and Science Policy.* New York: Praeger.

Killian, J.R., Jr. 1977. *Sputnik, Scientists and Eisenhower.* Cambridge, Mass.: MIT Press.

Kistiakowsky, George B. 1976. *A Scientist At the White House.* Cambridge, Mass.: Harvard University Press.

Lakoff, Sanford A. 1968. "The Third Culture: Science in Social Thought." In Lakoff (ed.), *Knowledge and Power.* New York: The Free Press, pages 1–61.

_____. 1977. "Scientists, Technologists and Political Power." In Ina Spiegel-Rösig and Derek de Solla Price (eds.), *Science, Technology and Society.* London and Beverly Hills: Sage Publications.

Landau, Martin, and Stout, Russell. 1979. "To Manage is not to Control." *Public Administration Review* 39: 148–156.

Leiby, James. 1978. *A History of Social Welfare and Social Work in the United States.* New York: Columbia University Press.

Long, T. Dixon, and Wright, Christopher. 1975. *Science Policies of Industrialized Nations.* New York: Praeger.

Mayntz, Renate. 1974. "Buerokratische Organisation und Verwaltung." In *Die Moderne Gesellschaft.* Freiburg: Herder. 2d. ed.

Meltsner, Arnold J. 1976. *Policy Analysts in the Bureaucracy.* Berkeley: University of California Press.

Mommsen, Wolfgang. 1959. *Max Weber und die deutsche Politik 1890–1920.* Tuebingen: J.C.B. Mohr.

Mosher, Frederick C. 1968. *Democracy and the Public Service.* New York: Oxford University Press.

National Academy of Sciences. 1973. *Report of the Committee on Research Advisory to the USDA.* Washington, D.C.

——. 1980. *Regulating Pesticides.* Washington, D.C.

National Science Foundation. 1980. *The Five-Year Outlook. Problems Opportunities and Constraints in Science and Technology.* 2 volumes. Washington, D.C.

Nelkin, Dorothy. 1977. *Technological Decisions and Democracy: European Experiments in Public Participation.* Beverly Hills and London: Sage Publications.

——. 1980. "Science and Technology and the Democratic Process." In National Science Foundation, *The Five-Year Outlook.* Volume 2. Washington, D.C.

Nelkin, Dorothy, and Pollack Michael. 1979. "Public Participation in Technological Decisions. Reality or Grand Illusion?" *Technology Review* (August/September 1979): 55–64.

OECD. 1972–74. *The Research System.* 3 volumes. Paris.

——. 1980. *Technical Change and Economic Policy: Science and Technology in the New Economic and Social Context.* Paris.

Oleson, Alexandra, and Voss, John, eds. 1979. *The Organization of Knowledge in Modern America, 1860–1920.* Baltimore: Johns Hopkins University Press.

Ostrom, Vincent. 1973. *The Intellectual Crisis in American Public Administration.* University, Alabama: The University of Alabama Press.

Parsons, Talcott. 1937. *The Structure of Social Action.* New York: McGraw-Hill. Free Press edition, 1949.

Pavitt, K., and Walker, W. 1976. "Government Policies Towards Industrial Innovation: A Review." *Research Policy* 5: 11–97.

Porter, David O. In preparation. "Accounting for Discretion and Local Environments in Social Experimentation and Program Administration: Some Proposed Alternative Conceptualizations." In Katharine L. Bradbury and Anthony Downs (eds.), *Do Housing Allowances Work?* Washington, D.C.: The Brookings Institution.

Porter, David O., and Warner David C. 1979. "Organizations and Implementation Structures: A Discussion Paper." Berlin: International Institute of Management.

President's Science Advisory Committee. 1965. *Restoring the Quality of Our Environment.* Washington, D.C.: Government Printing Office.

Price, Don K. 1954, *Government and Science.* New York: Oxford University Press.

——. 1965. *The Scientific Estate.* Cambride, Mass.: The Belknap Press of Harvard University Press.

Public Admnistration Review. 1977. Symposium on Policy Analysis in Government. PAR 37.

Rheinstein, Max, ed. 1954. *Max Weber on Law in Economy and Society.* New York: Simon and Schuster, 1967. First published 1954.

Rein, M., and White, S.H. 1977. "Can Policy Research Help Policy?" *The Public Interest* 49: 119–136

Rosen, S., ed. 1973. *Testing the Theory of the Military-Industrial Complex.* Lexington, Mass.: Lexington Books, D.C. Heath.

Schmandt, Jurgen. 1966. "The Scientific Statesman." *Les Etudes Philosophiques* 21: 165–186.

_____ . 1975. "United States Science Policy in Transition." In T. Dixon Long and Christopher Wright, *Science Policies of Industrialized Nations*. New York: Praeger.

_____ . 1977. "Science Policy: One Step Forward, Two Steps Back." In Joseph Haberer (ed.), *Science and Technology Policy*. Lexington, Mass.: Lexington Books, D.C. Heath.

_____ . 1978. "Science and Technology as Objects of Social Science Research." *Society for Social Studies of Science* 3: 14–22.

_____ . 1981. "Wissensformen der Politik." *Zeitschrift für Politik* (Munich).

Schmandt, Jurgen; Shorey RoseAnn; Kinch, Lilas; et al. 1980. *Nutrition Policy in Transition*. Lexington, Mass.: Lexington Books, D.C. Heath.

Simon, Herbert A. 1960. *The New Science of Management Decision*. New York and Evanston: Harper & Row.

Smith, Bruce L.R. 1966. *The Rand Corporation. Case Study of a Nonprofit Advisory Corporation*. Cambridge, Mass.: Harvard University Press.

_____ . 1977. "The Non-Government Policy Analysis Organization." *Public Administration Review* 37: 253–258.

Smith, Bruce L.R., and Karlesky Joseph J. 1977. *The State of Academic Science: The Universities in the Nation's Research Effort*. New York: Change Magazine Press.

Snow, C. P. 1961. *Science and Government*. Cambridge, Mass.: Harvard University Press. 2d edition with important post-scriptum, 1962.

Sohm, Rudolph. 1911. *Institutionen: Geschichte und System des Römischen Privatrechts*. 14th ed. Leipzig: Duncker and Humbolt.

Spiegel Roesing, de Solla Price, and Ina and Derek. 1977. *Science, Technology and Society*. London and Beverly Hills: Sage Publications.

Stinchcombe, Arthur L. 1959–60. "Bureaucratic and Craft Administration of Production: A Comparative Study." *Administration Science Quarterly*. 4:168–187.

U.S. Department of Health, Education and Welfare. 1976. *The Measure of Poverty*. Washington, D.C.: Government Printing Office.

Weber, Max. 1976. *Wirtschaft und Gesellschaft. Grundriss der Verstehenden Soziologie*. 5th ed. Tuebingen: J.C.B. Mohr, 3 volumes.

_____ . 1968. *Economy and Society*. Guenther Roth and Claus Wittich, eds. 3 volumes. New York.

_____ . 1966. *Staatssoziologie. Soziologie der rationalen Staatsanstalt und der modernen politischen Parteien und Parlamente*. Johannes Winckelmann, ed. Berlin: Duncker und Humblot. 2d edition.

Weiss, Carol. 1977. "Research for Policy's Sake: The Enlightenment Function of Social Research." *Policy Analysis*.

Weiss, Carol, and Barton, Allen H. eds. 1980. *Making Bureaucracies Work*. London and Beverly Hills: Sage Publications.

Wiesner, Jerome B., and York, Herbert F. 1964. "National Security and the Nuclear Test Ban." *Scientific American* 211 (October 1964): 27–35.

Wolf, Erik. 1939. *Grosse Rechtsdenker der deutschen Geistesgeschichte*. Tuebingen: J.C.B. Mohr.

3

Industrial and Technological Policies of Western* Economies

H. JEFFREY LEONARD and ROBERT GILPIN

Introduction

The tumultuous events of the 1970s have fundamentally changed the international economic environment. Outwardly, the decade will be remembered as one of rapid-fire shocks to the gradual consensus that had been building among the advanced industrial countries during the postwar decades of unparalleled economic expansion and prosperity. The abrupt ending of the Bretton Woods monetary system by President Richard Nixon on August 15, 1971, the almost overnight quadrupling of international oil prices, the sharpest and most concerted cyclical downturn to hit the industrial economies during the postwar era, and the adamant demands for a New International Economic Order by the developing countries, were but a few of the most prominent disruptions that characterized this transition.

But, in addition to raising these concerns, the difficult years of the 1970s exposed a long-term and less remarked upon problem within the key economies of the international system that is likely to hamper future world economic progress and political cooperation. That is, the prolonged slack in economic growth pointed out the existence or threat of structural weaknesses

*In this paper "Western" means western type economies and, therefore, includes the case of Japan.

in the industrial economy of nearly every Western nation. These structural problems have important international overtones and are liable to dampen further impulses for international cooperation and continuing international economic integraticn if considerable progress is not made toward their solution, especially in what is still by far the most important economy, the United States. Moreover, as will be noted below, the large-scale restructuring that may be necessary to ensure a new economic equilibrium within the United States and the international economy is encountering increased resistance from powerful interests in both labor and industry who want government assistance to preserve the status quo.

Within such a context, the United States and every other industrialized Western country has devoted increasing attention to developing explicit strategies to deal with structural problems in its industrial sector. These so-called industrial policies have centered around two major and frequently conflicting goals: (1) the promotion of technological development and industrial innovation in advancing sectors of the economy, and (2) the cushioning of decline in a number of lagging industries so as to preserve employment and to help these industries adapt gradually to changing world economic circumstances. This paper explores these two sides of Western industrial policies with the purpose of outlining the major concerns that should inform the expanding debate about industrial policy in the United States and the domestic and international constraints that surround such a debate.

The Growing Importance of Industrial Policy as an Instrument of the State

It is important to begin the discussion of current industrial problems and policies in the West by backtracking. The global events of the 1970s did not create the structural imbalances now apparent in the advanced industrial economies. Nor did these particular structural problems bring into existence the concept of industrial policy — broadly conceived here as government actions (or non-actions) designed to influence overall industrial structure or the structure of particular industries by halting, slowing, facilitating, or inducing adaptation to world economic developments.[1]

Since the end of World War II, the world economy has been undergoing a major transformation. Many varied factors underlie the changed international economy, but the effects of the contemporary technological revolution upon economic and commercial activities have been of primary importance. Of present-day technological developments and their economic effects, the most noteworthy are advances in air and sea transportation, which have decreased transportation costs and travel time between continents; improvements in radio, telephone, and television communications, which in conjunction with advances in transportation have facilitated the emergence of a global market; and, most significant of all, the unprecedented innovation of new products and of cost-reducing industrial processes, which has profoundly altered the relationship between technology and economics.

Three major interrelated economic consequences have flowed from these developments. All three have contributed to the growing need for the state to develop policies to guide the pace of development and change in the industrial sector. The first consequence, as many observers have pointed out, is the increased interdependence among national economies and the resulting greater sensitivity that each economy now exhibits to external changes in economic conditions. The high level of interpenetration that has been fostered among Western economies has significantly narrowed the range of economic policy options open to governments. The large-scale reductions in tariff barriers in many areas of trade that have come about under the General Agreement on Tariffs and Trade (GATT) have deprived governments (to a large extent at least) of what has historically been their major weapon for protecting domestic industries from foreign competition. In light of these tariff reductions, the importance and prevalence of so-called non-tariff barriers have grown. Many of the industrial policy initiatives being taken by governments to shore up lagging domestic industries fall into this broad category. Thus, whether purposefully or not, some industrial policy measures have taken over functions formerly performed by commercial policy. In addition, the lowering of trade barriers has increased the pressures on governments to try to come up with better plans for adaptation, as the OECD's Rey Report noted several years ago:

> Thus the process of freeing trade would seem to be increasingly bound up with other aspects of economic and social policy, notably a conscious effort to restructure activities and achieve a better adapted labour force.[2]

The general opening up of national economies has also stimulated more concern for industrial policy by undermining some of the once basic tenets of national economic planning and management. Charles-Albert Michalet has taken note of this in France:

> . . . the position of the state has been modified by the opening of the French economy. That opening has limited the state's means of control to various forms of macroeconomic regulation, and it has loosened the state's hold on the enterprises themselves. The emphasis on the primacy of macroeconomic coherence, which originally lay at the heart of the French indicative plan, has ceased to be accepted as valid.[3]

In France and in other countries, the state has turned increasingly to the use of structural policies in an effort to regain some of the national economic autonomy lost to international interdependence. But, as Göran Ohlin has pointed out, even though these new industrial policies have dethroned national planning, they have tended, to date, to be much more ad hoc in nature. Indeed, he says, one of the reasons that the non-tariff barrier type of industrial policies may be much more difficult to negotiate over than tariff barriers is precisely the fact that their expanded use more often than not represents "the triumph of the ad hoc."[4]

In addition to decreasing domestic autonomy, however, increased interdependence of national economies and the existence of international organizational forums, especially·the OECD, have helped to stifle governments from resorting to some of the most blatantly mercantilist policies of yesteryear. But these limits on what William Diebold calls "simplistic mercantilism,"[5] have also focused more attention on the need for government industrial policies. For example, an alternative to the exportation of unemployment through exchange-rate manipulation is the structural policy of propping up lagging domestic industries to maintain or increase levels of employment.

The second consequence is the enhanced role of technological innovation in economic growth and competition. International and domestic developments have forced the world's industrial nations, and especially the major trading nations, to give ever closer attention to their scientific and technological capabilities. The result in many countries has been a series of moves designed to improve both the training of scientists and engineers and the organization of science and technology and to make more efficient allocation of resources for research and development. The technological revolution that has engulfed the industrial world is requiring nations to develop conscious and systematic scientific and technological strategies.

To advance or even to maintain its position as an economic power in the world today, a nation must formulate long-range technological strategies to deal with certain unavoidable problems. In the first place, very considerable funds are frequently required to develop even the most elementary scientific and technological capabilities, and, of course, the investments required to develop certain areas of technology are enormous. Even the most wealthy nations must select for support specific fields of science and technology instead of attempting to compete across the board. Second, the long lead time between the conception of a new technology and the production stage means long-range planning and commitments.[6] Third, the rapidly growing appreciation of the impact of technology on human welfare and the physical environment is forcing governments to assess and to attempt to control the consequences of technological advance.[7]

Economists are now, for the first time, acknowledging that differential rates of technological innovation and diffusion of new knowledge are major determinants of the patterns of international trade. As a result of this fact, the theory of international trade has undergone a reorientation as dynamic elements displace the older emphasis on static factor endowments — land, labor, and capital.[8]

Thus, the international economy has become characterized by the increasingly rapid obsolescence of technologies and the undermining of traditional comparative advantages. Easily transferable technology, rapid transportation and communication, and high rates of capital and managerial mobility have given the notion of comparative advantage a much more fluid nature — it is constantly shifting, difficult to hold back, and sometimes subject to wild fluctuations.[9]

As industry becomes more science-based and innovation-oriented, the tradition and experience of a firm, or indeed of a nation, are of decreasing utility in the maintenance of compartative advantage. Even the possession of raw materials confers less of an advantage than in the past, owing to the rapid decline in the cost of marine transportatin. In fact, today a country deficient in raw materials, like Japan, can become a major industrial power, whereas a country like Sweden must worry over the declining importance of its iron ore resources.

In this environment, it is not surprising that governments have felt the need to develop better industrial policies that seek to promote technological development and to slow or smooth out rapid changes in the international economy. Even setting aside the mercantilist impulses that may underlie some efforts in this area, the most altruistic and enlightened government is bound to have an interest in trying to introduce a measure of structural continuity and predictability into the process of shifting international comparative advantage.

The third phenomenon that has flowed from changes in the international economy over the last 30 years has been the rapid expansion in the internationalization of capital and production. In particular, there has been a large movement abroad by American (and subsequently other national) corporations equipped to take advantage of the new conditions in the world economy and increasingly placing organizational emphasis on a worldwide production strategy. At the same time, however, national governments have had to be concerned with maintaining a balanced industrial base within their own borders. An important charge of industrial policies under these circumstances is to ensure that the domestic economy continues to sustain those industrial facilities that the state deems necessary for strategic, economic, or even social reasons. In the past, when it was in a much stronger competitive position *vis-a-vis* the rest of the world, the United States devoted little attention to this concern. However, as many observers have pointed out in recent years, it is now imperative that American policy makers carefully evaluate how the continuing location abroad of American corporations affects the U.S. economy and its industrial base.[10]

In addition to these three changes in the international economy, there have also been significant political and economic transformations at the domestic level during the past several decades that have increased the need for industrial strategies. The population explosion and the revolution of rising expectations have made economic growth a major goal of governmental policy in both the industrialized and the lesser developed nations. For the poorer nations of the world, expansion of the gross national product is necessary for survival; for the developed nations, it has become the primary means of ensuring full employment and of alleviating domestic discontent. Underlying the success of the contemporary welfare state is the fact that inter-group conflict over the just distribution of wealth has been displaced to a considerable degree by intergroup cooperation in the achievement of a higher growth rate. This substitution of a politics of growth for one of distribution has considerably changed the role of the state in economic affairs.

Furthermore, the growing need for the state to develop programs for planned structural interventions comes partly as a response to changing domestic political expectations that are seen occurring in all modern political systems. The state has been bombarded with rapidly rising numbers and types of demands from special interest groups — regional, professional, ethnic, public interest, business, and labor. The state is being called upon to provide for job security, more social services, better income distribution, regional development, social security and welfare payments, environmental quality, public health, and numerous other social welfare concerns. In one way or another, all these amount to demands on the state to correct the ills of past industrial growth, to assuage the workers and owners threatened by ongoing industrial change, or to clear away the bottlenecks seen to inhibit future industrial progress. The state has had to step in to make short-term structural interventions to meet many of these welfare demands; it has had to slow change so as to protect jobs, sectors, regions, and the environment. At the same time, the proliferation of short-term industrial policy measures of this sort have increased the need for the state to assess their impact on efficiency, structural adaptability, and growth, and to attempt to formulate longer term policies that ensure these goals are not vitiated. That is, industrial policy measures have increased both in response to social welfare demands and to the realization that these responses often leave long-term needs unattended.[11]

Though the turmoil of the 1970s did not create these circumstances, it did help to turn the resultant imbalances and structural weaknesses in the industrialized economies into much more immediate and menacing concerns — and thus focused more attention on the need for concerted industrial policies. For example, while rapid economic growth in the 1960s had helped to mask the problems of declining and noncompetitive industries by providing more employment in other sectors, the downturn in the overall economy has presented something of a scissors dilemma to national governments with regard to these declining industries. Governments are at once faced with the need to rely more heavily on declining industries to maintain desired levels of employment, and the need to make these industries more competitive in a situation of growing world surplus capacity. The prescribed action for dealing with one aspect of the dilemma was the worst possible medicine for the other. This sort of paradox, in turn, has complicated cyclical economic difficulties and helped to underline the fact that the economic maladies facing the West require the application of more than traditional macroeconomic measures designed to stimulate overall economic demand or expansion.

The 1970s created a more urgent climate under which Western governments made more concerted, if halting, efforts to create or stifle structural changes in the industrial system in accordance with domestic economic and political goals. Over the last decade, concern has magnified in all industrialized nations that they will be left behind or unable to maintain their competitive position unless they formulate appropriate strategies to adapt their economies to the imperatives of economic growth and competition.

Industrial policy has also been forced to shoulder the burden as a key governmental policy instrument for creating a "buffer zone" between the domestic economy and the harsh Darwinist forces at work in the interdependent and increasingly open international economic system. In short, as other domestic policy options have been circumscribed (e.g., tariffs) as the forces of economic and technological change have increased in velocity, and as domestic reactions to economic discomfort and dislocation have grown, the structural interventions taken by the Western state have been pushed much more toward the center stage of economic policy making. Governments are being forced to take more overt actions to shape their industrial structure, a task that previously was more likely to be left to market forces or handled through indirect measures (e.g., military strength, commercial and monetary policies, macro-economic interventions). Against this background, the next sections consider these industrial policies in more detail.

Industrial Policies in the West

Industrial policy has become more important in influencing both a state's ability to meet the domestic demands placed upon it *and* the conditions underlying its competitive intercourse with other nations. It is likely that the burdens placed upon industrial policy by the different factors described in the previous section are going to intensify in the future. As a result, most Western governments are, to one degree or another, struggling to come to grips with the conflicting demands and heightened responsibilities of industrial policy. They are making greater efforts to find the necessary balance of policies that induce, slow, and prevent structural changes — a balance that preserves enough of the status quo to satisfy domestic social welfare concerns and politically powerful vested interests, but also encourages enough flexibility and change to ensure the country's long-term competitivenss in emerging and "spearhead" fields of production.

At the same time, the industrial policies adopted by the state may have serious implications for other states as well as the enterprises and individuals in other countries. Quite frequently, this aspect of the state's industrial policy goes unbalanced, but there is some irreducible minimum below which ignorance of external impacts will induce more disorder in its international relations than the state considers desirable. Hence, within the broad confines of the "guided anarchy" that still prevails among them, the Western states are groping to pursue more coherent and consistent structural policies.

This pursuit, however, is to a large extent, fraught with contradiction. The conflicting nature of the demands placed on industrial policy make it almost inevitable that policies themselves will work in different directions,. Governments, as Jacques Maisonrouge, President of IBM Europe, points out, are "caught between a rock and a hard place."[12] On the one hand, their ability to compete in the international market depends on constant innovation and adaptation and a high level of mobility of the factors of production between sectors and locations. On the other hand, the demands of regional policies, the political volatility of unemployment, the political clout of entrenched industrial interests facing obsolescence, and numerous other domestic concerns strongly resist many of the requisites of international competitiveness — innovation, rationalization, mechanization, mobility, etc. Further, nearly every action taken by the state to encourage preservation, adaptation, or innovation in the industrial sector is bound to be seen by some foreign groups as tantamount to the provision of an artificial advantage in the international market.

Thus, under the rubric of a state's industrial policy, there is liable to be a constant tension between the state's welfare goals, its international competitiveness goals, and its international relations goals; it is unlikely today that one policy would work toward maximizing all of these goals.

Some Western states — most notably Japan and France — have come quite far in debating, explicitly outlining, and trying to effect industrial policies that do satisfy all the demands placed upon them. Yet, tension among the three broad goals of industrial policy is inevitable even in these nations. Japan, which has been highly effective in encouraging adaptiveness and innovation in a dynamic economic setting, still probably faces the least difficulty of the major Western nations in balancing its domestic welfare and international competitiveness goals. But the measures that have helped Japan to satisfy these have taken a heavy toll on the cordiality of Japan's relationships with its trading partners, who feel that Japanese industrial policies succeed at foreign expense.[13] French structural policy, in spite of the high priority attached to it, is bogged down in something of a stand-off. For all intents, France is pursuing two separate and conflicting sets of industrial policies — those to stimulate advanced industries, and those to meet welfare needs — and is facing serious problems in efficiently accomplishing either. In France today, industrial policy is, as Lionel Stoleru said a decade ago, largely a matter of the state giving with one hand and taking away with the other.[14]

The concept of a single coherent national industrial policy, if somewhat contradictory in Europe, is hardly even grasped as a legitimate state pursuit in the United States. Indeed, there are many who would argue the opposite; the idea of the state having one industrial policy is still viewed suspiciously in the United States. This is strongly evidenced in the OECD's volume on *United States Industrial Policies,* which underlines the fact that "the American philosophy of economic progress has never called for a 'co-ordinated' government attempt to intervene in the operations of various industries."[15] Instead, the report emphasizes that intervention "has normally been *ad hoc*, to deal with particular situations."

Still, in the United States, as in Europe and Japan, a number of government policies have long existed in the areas of primary concern for Western Industrial policies: (1) strategies for technological innovation, and (2) assistance to lagging industries, either to prop them up or to help them to adapt gradually to rapidly changing international economic circumstances. A review of how industrial policies in Europe aand Japan are addressing these two concerns and some of the problems which may arise in the future follows:

Strategies for Technological Innovation

A nation can follow one of three basic strategies with regard to governmental efforts to stimulate continuing technological innovation. The first is to support scientific and technological development across the broadest front possible. A nation following this broad front strategy seeks to maintain a position in all the advanced fields considered to be of military, economic, and political importance — especially atomic energy, computers, and aerospace. This strategy is, of course, the one followed by the two great powers, the United States and the Soviet Union. It is also the strategy that Great Britain attempted to follow until the early 1960s and that France pursued under the leadership of President Charles de Gaulle.[16] This strategy has been undergoing a critical review in France since the beginning of the 1970s.[17]

The second strategy is scientific and technological specialization. The essence of this strategy is to select for support specific areas of science and technology, usually of commercial utility, and to concentrate one's resources upon them. In the selected areas, an innovative policy is required at every stage, from basic research through technological development. Among the countries that have most successfully followed this strategy are Sweden, the Netherlands, Switzerland, and, increasingly, Great Britain.

In contrast to the first two strategies, which are innovative, the third is an adaptation strategy. Emphasis is placed on importing foreign technology by the purchase of licenses; in the period since World War II, imported technology has most frequently been American. Although this strategy, like the second, implies specialization, it differs in that relatively little basic research is carried out. Instead, indigenous scientific and technological resources are concentrated on improving and redesigning imported technologies, especially for subsequent export. The country that has perfected this strategy is, of course, Japan.[18] In addition, West Germany has also relied heavily upon the importance of American technology in advanced fields such as computers, atomic energy, and aviation.

A nation's choice of strategy reflects its social, economic, and security circumstances and objectives; therefore, the choice of strategy cannot be separated from the nation's broader domestic and foreign policies. Differing national strategies can be identified, but it must be noted that they are not exclusive strategies, but rather they characterize the major emphasis of a country's scientific and technological policies. For instance, the United States obviously cannot do everything; Sweden does imitate in some areas; and Japanese firms, for their part, obviously do innovate. Furthermore, each strategy entails its own peculiar risks and opportunities, and, as the cases of France and Japan attest, the consequences of choosing the right or wrong strategy can be momentous. Indeed, a nation's strategy may change as its circumstances, capabilities, and objectives change over time.

Looking back over the postwar era, there are two societies which provide very clear models of what to do and, perhaps of greater importance, what not to do with respect to technological innovation. On the one hand, the Japanese have pursued an exceptionally successful policy for technological development since the end of the Second World War. While the Japanese economic and political structure differs significantly from that of the United States, the Japanese experience still has relevance for the United States. On the other hand, the British have made many serious mistakes in the area of technological innovation over the past several decades. Unfortunately, American policies have been closer to the British than to the Japanese model.

Although the experiences of other societies in the area of government support for technological innovation have limited utility for the United States, they do suggest pitfalls to be avoided and policies that have worked with great success. The lessons to be derived from a quick look at British and Japanese experience, for example, are especially important for two reasons. In the first place, pressures are presently developing rather rapidly in the United States to pursue courses of action that have been serious failures abroad. Secondly, the United States situation with respect to technology is now such that policies that have worked very well in particular countries have an increased relevance for the United States.

Japan's technological strategy has been characterized as one that exploits a variety of "technological niches."[19] Under the heavyhanded guidance of the Ministry of International Trade and Industry (MITI) and with careful analyses of potential export markets, the Japanese have purchased from abroad the technologies through which they have achieved their impressive rate of economic growth and export trade. To achieve their economic conquests, they have, in effect, employed the same techniques of close government-industrial cooperation, individual discipline, and unswerving dedication to the objective that once enabled their military conquests.

The Japanese have been able to concentrate their energies and resources on commercial technological development because of several highly favorable circumstances. In the first place, their alliance with the United States has relieved them of heavy expenditures on defense. A remarkably small fraction of their total R & D has been defense-related. And, secondly, the United States has encouraged the diffusion of American advanced technology to Japan. Yet, the important fact is that the Japanese took advantage of this situation to do most of the right things.

In the first place, the Ministry of International Trade and Industry in conjunction with the banking system and the large trading houses has provided central guidance and leadership to the Japanese economic revival. At the highest levels of government, industry, and banking the Japanese have coupled and integrated economic and technological policies. Appropriately labeled, but frequently misunderstood, "Japan Incorporated" gave central direction to the implementation of Japan's policy decision to rebuild the Japanese economy upon the most advanced technologies and industries.[20]

The second most noteworthy feature of the Japanese experience is that they have successfully coupled economic and technological policies to achieve one overriding objective: rapid economic growth. Through various economic policy instruments, the Japanese government and particularly MITI created incentives and disincentives that forced Japanese industry to innovate and adopt progressively higher levels of technology. For example, contrary to the myth that Japan Incorporated means government protection of industrial firms, Japan's highly selective and discriminating tariff policy has been ruthless with respect to inefficient and low-technology firms. It has been geared to wipe out industries in low-technology and decreasingly competitive sectors (textiles, shoes, etc.), and to protect firms in higher technologies (electronics, automobile, etc.) until they are strong enough to meet foreign competition. Furthermore, in contrast to American industry, the Japanese have tended to keep their high-technology industry at home. Japanese foreign direct investment is mainly in extractive and low-technology industries, though this is changing due to Japanese concern over foreign trade barriers and environmental degradation and overcrowding in Japan.

Thirdly, the Japanese have stressed the demand rather than the supply side of technological innovation. Market demand and profitability have determined the allocation of R & D resources. As such, and again contrary to myth, the Japanese government has played almost no direct role in technological innovation. In contrast to the United States, most of the funds of R & D are provided by industry itself. The major contribution of the government has come through a heavy investment in education and the provision to industry of a highly skilled labor force, including scientists and engineers. In addition, the government has financed applied research and experimental development in areas of commercial importance (shipbuilding, for example). The commercial development of technology has been left up to industry and its assessment of user demand. The Japanese government through its overall economic policies has managed demand and the economic environment, but it has left to private industry the task of bringing a technology to the state of commercial development.

In contrast to the United States, the Japanese have emphasized the adoption of foreign technologies and applied R & D in consumer technologies. They have given relatively little attention either to basic research or to big, prestige technologies. The greatest gains to Japanese economic growth and productivity have come, in fact, through the enhancement of intermediate products (materials, supplies, etc.). The Japanese success in importing and adapting technology has been due to the development of a strong R & D capacity within Japanese industry. Through the employment of large numbers of scientists and engineers, Japanese industry has been able to monitor, select, and adapt foreign technologies. In general, these professionals have performed adaptive, rather than original, research.

The latest example of remarkable success for Japan's technological strategy appears to have come in the burgeoning microchip industry. Although American companies were largely responsible for advancing microchip technology to the stage where it could be mass produced cheaply, Japanese companies were quick to jump into this market. Thus, in 1978 and 1979, when demand skyrocketed for microchips to supply numerous consumer electronics products, Japanese companies captured a large segment of the American market.

Nevertheless, among students of Japanese economic and technology policies, a debate has taken place over the past several years regarding the feasibility of the Japanese formula in the future. Certainly, two of its fundamental ingredients are changing. In the future, the Japanese will have to invest more of their economic and technical resources in defense. Additionally, as their economy has come abreast of the American economy and as American firms become more restrictive about licensing their technology to the Japanese, Japan will be less able to depend on importing foreign technology. For these reasons, one will undoubtedly see a greater Japanese emphasis on basic science and technological innovation.[21] The issue, which only the future can decide, is how well the Japanese can make this jump from successful adoption to technological innovation.

It is undoubtedly a misnomer to speak of a British technological strategy.[22] Unlike the Japanese, British economic and technology policies have had little clear direction. Cooperation among the government, financial, and industrial communities has been poor. Many policy and institutional initiatives have been tried, failed, and eventually abandoned. Economic and technology policies have frequently been at cross purposes rather than reinforcing one another. On the one hand, the British have sought to rebuild an industrial and economic structure which had been tied to an imperial system. On the other hand, British economic policy has emphasized redistribution of wealth and the nationalization of industry at the same time that they have recognized the need to create large, efficient industrial entities to compete against the Americans, Germans, and Japanese.

Underlying the British malaise has been the problem of making the adjustment from the status of a global imperial power to that of a middle-sized European state. The emphasis and errors of British technological policies can be attributed in large measure to the economic and psychological difficulties of making this adjustment. While this unique circumstance would seemingly make the British experience irrelevant to the United States, the contrary is the case. As we pointed out earlier, the United States today faces in part (though fortunately to a much lesser degree) the problem that has long faced the British. The relative industrial decline of the United States poses a similar challenge to us. Moreover, in our effort to meet this challenge, we face the same temptations to make the same errors that the British have committed.

The experience of Great Britain in technological innovation presents a curious paradox. At the same time that many of the most important and remarkable technological innovations of the past several decades have British origins — jet propulsion, holography, hovercraft, radar, etc. — the British have failed to take full advantage of these innovations and, on the whole, have made very poor use of their rich scientific and technological resources. The overall failure of British technological policy has been a continuing source of frustration for successive British governments. Despite the fact that government policy in Great Britain has been to make government-financed R & D more relevant to industry, the British have failed to meet this objective.

The reasons for these failures obviously lie deep in British society and economic organization. Yet, several aspects of British policy for the support of technology appear especially important. Moreover, British experience holds noteworthy lessons for the United States because of increasing pressures in this society to move precisely in the direction which has proven so unfortunate for the British. As in Great Britain, there are calls in the United States for disguised subsidization of industry under the subterfuge of security of supply, the scale of technology, and so forth.

In the first place, British Government expenditures like those of the United States have been overly concentrated in a relatively few high technology areas, such as defense, space, and atomic energy. The goverment has taken upon itself the role of entrepreneurship and has concentrated upon commercial development instead of on research, exploratory development, and related activities. In substituting its judgment for that of private entrepreneurs with respect to the commercial "ripeness" of particular high technology projects, the government has assumed a responsibility and task which governments do not do well. As a consequence of this neglect of the market, very few of these costly projects have had commercial success.

Secondly, and directly related to the first weakness, the British Government in a number of significant cases has made commitments to full-scale commercial development of particular technologies too early and on too big a scale. The British Government has failed to adopt the wiser policy of an incremental, step-by-step approach to R & D policy. At the same time, too little has been spent on the necessary applied research, project planning, and exploratory development that should precede the commitment to large scale development. As a consequence, there has been a neglect of more traditional sectors of the economy which, for historical and institutional reasons, tend to under-invest in R & D.

Thirdly, the British have failed to integrate sufficiently the three estates of science and technology: universities, government, and industry. They have failed to create the necessary mechanisms to bring together the sources of new scientific-technical knowledge and the industrial utilizers of knowledge. A disproportionately high fraction of British R & D has been conducted in government laboratories or in industry-wide cooperative laboratories catering to specific industrial sectors (steel-making, machinery, textiles). While this latter set-up has served to improve the state of the several technical arts, as the so-called Rothschild Report points out, the spill-over of government-supported military and related research into the private industrial sector has been minimized.[23]

In a sense, the underlying failure of British policy for R & D has been that the government has tried to supplement, rather than complement, the private market. Unlike the Japanese, they have failed to couple economic and technology policies. As a consequence, although the British are among the most technologically rich and resourceful people in the world, they have been unable to harness these resources to generate a sufficiently high rate of economic growth and competitive exports.

The British experience illustrates one last point which should be underscored. While the extent of government support of civilian, industry-related R & D varies in industrial economies, it is curious to note that the level of government support for commercial development is inversely correlated with the technological prowess of the economy. In addition to Great Britain, the two industrial economies with a relatively high level of government support — France and the Soviet Union — tend to trail those economies where the government's role in big technology has been least: Japan, Sweden, The Netherlands, Switzerland, and the Federal Republic of Germany. Although one cannot draw the conclusion that government support of big technology is harmful to R & D, this negative relationship at least suggests that government support in itself is not the solution to a declining technological base.

Aid to Declining Industries

In addition to delineating strategies for stimulating technological innovation, most OECD governments have tried to serve their goals of economic efficiency and international competitiveness by developing elaborate policies designed to encourage positive structural changes in, or a shift in, resources away from lagging industries. But in actuality, these efficiency- and competitiveness-oriented goals have quite frequently been overridden by domestic political concerns. To a large extent, the short-term orientations of most Western governments and the parochial interests of important constituency groups have often encouraged primary emphasis of industrial policy responses to be placed on preserving the status quo, or at least slowing structural changes in many industries.

There can be little doubt that many government measures that have been subsumed under the heading of industrial or structural policy in recent years have been heavily weighted toward inhibiting change. As William Diebold has recently written: The expressed aim of industrial policies may well be "to create strong new industries which will be competitive in world markets but much more often industrial policies are efforts to shore up old and weak parts of the economy or, at best, to tide them over a period of transition."[24]

German industrial policies of the early 1970s illustrate the point. Broadly speaking, these policies had three aims: The first aim was to stabilize and improve the income to employers and employees in the lagging industries.

The second was to help industries adapt to external change, which in practice meant slowing the decline of old industry more than accelerating the growth of new industry. The third aim, which was receiving a great deal of public attention at the time, was that of promoting research and development in the high technology industries. But, as George H. Kuster has noted, the weight assigned each of these objectives was far from equal. He writes:

> The political priority of the respective aims can be adduced from the fact that in 1970, 66 percent of the total amount of Federal subsidies and tax allowances was spent on the aims of conserving declining industries, 28.5 percent on easing the adaptation of industries, and only 5.5 percent on the aims of boosting productivity.[25]

West Germany is hardly alone. Despite official proclamations to the contrary, it is difficult to escape the conclusion that industrial policies in most OECD countries are heavily biased toward protecting domestic industries by helping them maintain existing structures or retarding the rate of decline. Even large-scale government efforts to build up national champions in the scientific industries do not offset the resources being expended by governments to hold back structural change. Moreover, all signs point to more, not less, government support for declining industries in the future.

Thus, although industrial policy in Europe has blossomed as a response to the perceived need to catch up to America and to stimulate technological innovation, it is in danger of evolving into something altogether different. Though relatively successful in stimulating certain positive structural changes and in encouraging ongoing research and development, most European industrial policies have also become major instruments for achieving state welfare goals in recent years. While this may be viewed as a positive development by the beneficiaries of those structural policies, which serve welfare goals, many of these actions do more to stifle or slow structural change than to induce it.

Many of the change-retarding actions being pursued under the guise of structural policies have become very expensive propositions and are now deeply imbedded in the fabric of society. Regional policies designed to encourage industrial location in lagging regions have drained funds from more innovative-type industrial policy coffers at an ever-increasing rate, but this has had little or no positive effect in terms of encouraging industrial adaptivity or innovation. In Britain — even with the Thatcher government's announced cutbacks of $530 million over three years — and France, regional industrial aid programs now cost well over a billion dollars per year. Italy shovels out an average of two billion dollars per year for regional aid. Even West Germany offered nearly 900 million dollars in cash and special tax breaks for 1979 to new investors in certain areas, particularly those in the lagging eastern border regions.[26]

The pursuit of national prestige and technological independence, while potentially encouraging of adaptability, has often created problems and bottlenecks of its own, as John Zysman has shown in his study of the French electronics industry[27], and as the case of the British and French Concorde illustrates. Efforts at export promotion have not always generated competitive success. Great Britain, for example, has outspent every other Western nation in recent years to promote its exports with no positive results — unless the 1978 expenditure of one billion dollars of government money to enable British Steel to undersell even the efficient Japanese producers on the world market can be called "positive."[28] Italy, according to Steven Warnecke, provides a classic case of the problems created by subsidies to industry. In effect, the country now has a "dual economy in which politically connected groups, often allied with the once innovative state holding companies, ENI, and IRI, drain resources from the rest of the economy."[29]

Agriculture in Europe is another, by now classical, example. The European countryside is filled with people being subsidized to remain on uneconomical farms. Social resistance to leaving these farms is overwhelming, and it would take a stronger state than even those of Europe to stop the flow of supports to unproductive farms and grapeless winemakers and to reduce the production of mountains of butter in the European Economic Community.

In summary, the already pressing difficulties of existing structural imbalances, supported by strong domestic constituencies, the prospects for continued unfavorable economic conditions, and the exigencies of welfare-state politics pose the very real danger of industrial policy reactions by governments that are largely competitive in the negative sense. Instead of focusing on spawning competition to shift resources and workers into new forms of production, Western industrial policies that react primarily to short-term demands are liable to concentrate more and more on shoring up lagging industries and, thereby, intensify cut-throat international competition to dispose of present products and stave off massive unemployment. Such a development is not only destined to exacerbate political conflicts among nations, it is likely to perpetuate lower levels of growth and worsen future economic hardships.

Jan Tumlir has characterized the tension between the short- and long-term interests of industrial policy nicely. "Economic progress," he writes, "implies novelty and social adaptation to it. But before this insight dawns upon the society, the state has been called upon, and has promised, to make possible progress without change."[30] But in such a situation, one nation's "progress" can only be secured by another's loss. As a consequence of such defensive industrial policy measures, the new emergence of industrial policy has been viewed in some circles as an indication of the West's rapid retreat toward protectionist economic policies. The unifying characteristic of most newly developed structural policy, says Göran Ohlin, is its "unabashed nationalism"; it is clear that "industrial policy is becoming increasingly noxious in the eyes of competing countries."[31]

The American Industrial Economy: Decline or Response?

In 1974, one of America's most distinguished and respected economists, Charles Kindleberger, published an article in which he considered whether or not the American economy was undergoing a climacteric. Like Great Britain in the latter part of the nineteenth century, the United States was being overtaken and surpassed by more dynamic foreign economies, particularly those of West Germany and Japan. The evidence was mounting, he argued, that the United States was not only falling behind in overall economic growth, but in the innovation of new technologies. Like the British, rather than stimulating the development of new products for world markets, many American government programs are designed to try to hold their own in older areas of technologies — such as steel and automobiles — and not doing very well at that.

In recent years, the flurry of debate concerning industrial policy in Europe and Japan, coupled with the growing perception that the United States is losing its once almost unassailable industrial prowess, have inevitably created more concern for developing an overall industrial policy in the United States. There is talk now of the importance of developing a coherent set of guidelines, even if not a centralized policy, by which to direct and evaluate the government's numerous ad hoc structural interventions. Pragmatic considerations in the realms of domestic and international relations are forcing on the United States the "need for purposive efforts moving toward an industrial policy . . . instead of relying on spillover tradition, and whatever built-in forces [that have] looked after American interests in these matters in the past."[32] National interest dictates that the American state, like its trading partners in the advanced industrial world, develop a clearer picture of the desirable structure of the national economy and better coordinate the policy tools that aim to accomplish this vision.

In some respects, the relative industrial decline of the United States over the last three decades was inevitable. At the end of World War II, the U.S. economy dwarfed all others in output and innovation. To some extent, this was an artificial situation, an advantage which even the most innovative and most strategic industrial policy could not have maintained. This is particularly true in light of the fact that the rebuilding of Europe and Japan through transfers of American technology, the dispersion of American managerial know-how, and huge doses of direct economic aid, was a major political goal of American foreign policy. Under these circumstances, it is doubtful whether the postwar knowledge and technology gap that supported American predominance over other Western economies could have been perpetuated for more than two or three decades. The reason for this relates to the whole matter of sequence and form in the international diffusion of knowledge and technology. Roughly, as Raymond Vernon has noted, this process proceeds in four stages:

1. the innovating country begins as sole producer and exporter of a new product,
2. foreign imitators begin production, decreasing the export their country accepts from the innovating country,
3. foreign imitators compete with the innovating country in global markets,
4. foreign imitators compete with innovating country in its domestic market.[33]

This cycle points out an inherent dynamic once significant product innovation has taken place: A leader emerges and other countries aspiring to "catch up" immediately begin devising strategies to do so.

Inevitably, as Thorstein Veblen noted earlier in this century, the innovating country must pay a price for having taken the lead. Observing the impending obsolescence of Britain's industrial plant and the rise of new European — especially German — industrial challengers, Veblen noted that even without her numerous mistakes and growing lethargy, Great Britain could probably not have beaten back the industrial imitators:

> All this does not mean that the British have sinned against the canons of technology. It is only that they are paying the penalty for having been thrown into the lead and having shown the way.[34]

It is in such a context that the rise of European and Japanese industrial policies must be viewed. A new cycle of international production began after World War II, with the United States as the innovating country and the world's preeminent producer in a number of new and leading sectors of production — synthetic materials, electronics goods, automobiles, and numerous other high-technology items. Economic, political and physical circumstances created by the war thrust the United States into an across-the-board technological lead over the rest of the world. The product cycle was simultaneously opened for a broad range of technologically advanced sectors and, what is more, the Americans assumed the lead in almost every one of these.

After the initial phase of rebuilding and economic recovery during the 1950s, European attention turned increasingly to the problem of reducing European dependence on American technologies in many advanced industries — the desire grew not only to speed up the dispersion of existing technology, but also to ensure that in the future Europe would not have to rely on the United States as the innovator of new technology. In essence, the strategies devised to accomplish these goals were efforts by European governments "to imitate the methods used in the United States to develop new technologies and spread them throughout their industries."[35] The policies adopted in Europe were primarily geared to recreate artificially what were considered to be the secrets of American success — management techniques, competition, mobility, decentralization, size of firms, etc. Obviously, these goals could not be realized through traditional macroeconomic interventions or national planning policies. More concerted programs of direct structural interventions were necessary in order to remake the European industrial sectors over in the American image.

Such were the goals that came to inform the growing concern with industrial policy in Europe during the 1960s. Early initiatives in this area concentrated on stimulating research and development planning, promoting mergers, rationalizing enterprises, reorganizing the credit and banking sectors needed for the development of industrial activities, and providing appropriate infrastructure for a modern industrial economy.[36] For Europe, these efforts at restructuring the industrial sector were necessitated by the belief that only through their successful implementation could they avoid the fate of being left behind every time the leader advanced its technology one more notch.

Today, the cycle of international productive dispersion, begun with America's innovative leadership in the leading postwar industries, has effectively run its course — at least insofar as the United States and its major economic rivals are concerned. In part, the challenges now posed by Europe and Japan to American technological and overall economic supremacy can be attributed to the successful structural measures taken to increase international competitiveness, facilitate the spread of knowledge and technology, and stimulate domestic innovation. Yet, the dynamics of the process that encourages the international diffusion of technology and knowledge make it difficult to argue that an effective industrial policy could have long defended America's disproportionate lead. Even if the transfer of knowledge and technology — and, therefore, the erosion of the American lead in advanced industries — was speeded by actions taken by subsidiaries of American multinational corporations and by the government itself, it is doubtful whether anything short of complete blockage of all knowledge transfers could have altered the cycle fundamentally. Furthermore, European and Japanese industries have now lept into the roles of innovators in certain industries, a development that cannot be blamed on the lack of defensive industrial policies in the United States. If the United States can be faulted in its failure to devise a strategy to reduce the "penalty of taking the lead," it is because it has not been able to innovate new products quickly enough to stay one step ahead of potential imitators, not because it failed to stop them from acquiring knowledge. But, this is no easily accomplished task, particularly in a world where the tools of scientific advancement are widely dispersed, relatively equalized, and intensely competitive. At any rate, looking to the future, even if the United States forges a strong industrial policy and leaps back into the saddle as chief innovator, it is doubtful whether the unique postwar circumstances can ever be duplicated — particularly, the synchronization of so many leading sector product cycles in time and under one national leader.

Still, the danger of further relative decline, and even the gradual shift of international innovation from the United States to countries such as West Germany and Japan, is very real today unless strategic government actions are taken. How can the United States respond to this challenge?

Response to Decline: Three Available Strategies

There are essentially three strategies that an economy can follow in response to its relative industrial decline.[37] Certainly no country has pursued, or probably could be expected to pursue, one of these three strategies to the exclusion of the others. But the relative emphasis on one strategy rather than another is of immense political and economic importance.

In the first place, such an economy can export capital in the form of loans or portfolio investments to other economies. It can in effect become a *rentier* and increasingly live off the earnings from its investments overseas. Because the profit rates tended to be higher abroad than at home and financiers controlled her investment capital, this strategy of portfolio investment was the one chosen by Great Britain in the latter part of the nineteenth century.

A similar strategy has been followed increasingly by the United States since the end of the Second World War and more particularly since 1958. The strategy of foreign direct investment, which has been emphasized by American multinational corporations, is a far more complex phenomenon than the earlier British strategy. In part, it, too, is motivated by a differential rate of return on capital that favors foreign over domestic investment. However, if interest payments and capital gains were the sole motivation of American foreign investors, the pattern of American foreign investment would not be fundamentally different from that of Great Britain. And, in fact, substantial amounts of American foreign investment is portfolio investment.

The fundamental differences between the American strategy of direct investment and the British strategy of portfolio investment derive from the fact that the primary American investors are corporations rather than financiers or bankers. For this reason, two other critical factors are involved in their decision to go multinational and to establish branch plants or subsidiaries overseas. These American multinational corporations seek to capture an additional rent on some oligopolistic advantage — a product or process innovation, a well-known trademark, or superior access to capital. The possibility of obtaining both a higher rate of profit than at home, plus "monopoly rents," was not available to British investors, or at least not to many of them. Moreover, the threatened loss of its monopolistic advantages and of market shares to foreign (or domestic) competitors is itself a further stimulus to foreign direct investment on the part of American corporations. For these several reasons, therefore, the "maturing" of the American economy has led to an immense outflow of capital in the form of direct investment.

In addition to this foreign investment strategy, there is, in theory at least, a second strategy available to the economy in response to its threatened relative industrial decline. This is the rejuvenation of the economy itself. In particular, this strategy implies the development of new technologies and industries and the redirecting of capital into neglected sectors of the economy. Through investment in research and development, for example, capital can be employed to innovate new products and industrial processes. In terms of foreign economic policy, this strategy implies a policy which emphasizes trade rather than foreign investment. For domestic and international political reasons, there is a sound basis for arguing that a greater emphasis on this strategy on the part of Great Britain in the nineteenth century would have served her national advantage. A similar argument can be made for the United States today.

There is yet a third strategy in response to relative decline: "reaction" is perhaps the most appropriate characterization. That is, an economy may withdraw into iself. As other economies advance, competition increases, and the terms of trade shift to its disadvantage, the declining economy retreats into protectionism or some sort of restricted preference system. It throws up barriers both to the export of capital and to the importation of foreign goods. It favors preferential commercial arrangements. Too little is done to reinvigorate its domestic industrial base. This tendency is certainly one that has long been at work in the case of Great Britain; it is increasingly evident in the United States.

Advanced industrial economies tend to be highly conservative and resistant to change; the generation of new products and production processes is an expensive one for industrial firms with heavy investment in existing plants; labor can be equally resistant. The propensity of corporations is to invest, in particular, industrial sectors or product lines, even though these areas may be declining. That is to say, the sectors are declining as theaters of innovation; they are no longer the leading sectors of industrial society. In response to rising foreign competition and relative decline, the tendency of corporations is to seek protection of their home market or new markets for old products abroad. Behind this structural rigidity is the fact that for any firm, its experience, existing real assets, and know-how dictate a relatively limited range of investment opportunities. Its instinctive reaction, therefore, is to protect what it has. There may be no powerful interests in the economy favoring a major shift of energy and resources into new industries and economic activities. In short, an economy's capacity to transform itself is increasingly limited as it advances in age.

Under these pressures, there is a very real possibility that industrial policy will increasingly become entrenched as a servant of welfare goals: preserving employment in uncompetitive sectors and redistributing wealth among people, sectors, and regions. These programs work to increase rigidity and to resist structural change, not to encourage it.

As was already noted in the last section, many of the industrial policies of certain European nations may be evolving more and more into this third strategy. This evolution does not appear to be one guided by conscious choices on the part of European governments as much as it appears to be a result of prevailing social attitudes and expectations about the state's authority and responsibilities.[38] In a word, the welfare state and the requirements being placed upon it to redistribute wealth and compensate those disdvantaged by change in industrial societies are much further advanced and more firmly entrenched in Europe than in the United States.

All things considered, the European nations face a major quandary for the future. Industrial policy has become a major instrument of the welfare state and, as such, is being diverted from its forward-looking mission to encourage flexibility, adaptation, and innovation. (Japan, for cultural reasons in addition to its continued growth, has, to date, been far more successful in meeting its domestic requirements for buffeting groups from change through positive rather than regressive stuctural measures.)[39] In Europe, adjustment assistance has often become adjustment resistance.[40] Temporary aid has become a euphemism for the maintenance of surplus industrial capacity. Some of these trends toward protective industrial policy measures can certainly be found in the United States. If the economic situation continues to be sluggish, demands for structural interventions to serve welfare goals are going to increase. But, while in Europe one may indeed ask whether the negative aspects of industrial policy are reversible,[41] there is some hope that the United States will be able to formulate a forward-looking industrial policy before succumbing to too many of the multiple welfare demands that work at cross purposes with the important long-term goals of industrial policy, for example, technological and economic rejuvenation. This is not so much an argument against welfare policies as it is one against making industrial polices a major instrument of welfare goals.

It is increasingly evident that American industrial policies must focus on a strategy of rejuvenation of its domestic industrial and economic base. In designing an industrial policy to meet the needs of America's future, the

American government must reevaluate its continuing emphasis on the overseas expansion of American corporations — not necessarily by banning such expansions, but certainly by assessing their impacts on the domestic economy and its industrial base, and by ending numerous government projects that have in the past actually encouraged and given financial banking to American companies to locate facilities abroad. Unlike the situation that prevailed following World War II, the economic and political foreign policy interests of the United States are no longer served by policies that encourage the dispersion of advanced American technologies and multinational corporations throughout the world.

On the other hand, the United States must strongly resist the opposing temptation to retreat into outright protectionism or to fall into the trap of providing ever-expanding financial assistance to prop up dying industries instead of devising programs that shift resources away from these industries and provide direct assistance to help move workers into more productive jobs.

All this makes it more and more evident that American industrial policies must be oriented around a strategy of rejuvenation of the domestic industrial and economic base. Regrettably, the rejuvenation of an economy and the shift of resources to new leading sectors would appear to be the consequence of catastrophe such as defeat in war or an economic crisis. It took near-defeat in the First World War for Great Britain to begin to restructure her economy, though even then she did not go far enough. It is not surprising today that two of the most dynamic industrial powers — Japan and West Germany — were defeated nations in the last World War.

Moreover, rejuvenation is going to be extremely difficult in light of worldwide economic and technological realities. There is good reason to believe, as Peter Drucker and others have argued, that the United States and its economic partners have exhausted many of the innovative possibilities of the industries upon which American economic power, foreign investment, and the immense growth of the last several decades have rested — the internal combustion engine, manmade fibers, electronics, and steel. The growth curves of these industries, much like those of cotton, coal, and iron in the last century, appear to have flattened out. They appear to have ceased to be major theaters of innovation and future industrial expansion, at least in the developed countries. Thus, with the apparent exhaustion of technological opportunities and the closing of the technology gap in the commercial (though nct military) technology among the industrial and the industrializing countries, economic conflict among industrial economies has greatly increased.

In the contemporary world, the shortening of the international product cycle due to the more rapid diffusion technology, the liberalization of trade, and the emergence of many industrial economies, have intensified competition. Profit margins have declined; investment and growth have slackened.

The industrial world could well enter a period of mercantilistic conflict similar to that which characterized the period prior to World War One and also the interwar period. As a consequence, though we may neither have reached the "limits to growth," as some doomsday prophets hold, nor be entering the severe depression phase of a Kontdratieff wave, as others argue, the world is certainly entering an era of major adjustment to new economic realities.

There is, in short, a tremendous bottleneck at the top of the world industrial ladder that is making things tougher and economic growth more difficult as everyone bunches up closer and closer behind the most advanced countries. Like it or not, and in spite of numerous safeguards, all but the most advanced technological innovations are destined to "trickle down" at a rapid rate.[42]

In light of these realities, what does a strategy of rejuvenation entail? First, although economic conflict has been intensified by the energy crisis, shortages of resources, and worldwide inflation, the critical issues of resources, environment, and inflation, viewed from a longer perspective, could have a beneficial effect. They may constitute the "catastrophe" that could stimulate a rejuvenation of the American economy. In the search for solutions to these pressing problems, the United States is being forced to innovate a new order of industrial technology and economic life. If this search leads to technological breakthroughs and the fashioning of a new international division of labor, we may yet escape the mercantilistic conflict and economic decline which threaten us.

A necessary ingredient in any rejuvenation strategy is an improved capacity in civilian-related research and development programs sponsored by the government. Although the U.S. government funds massive amounts of research and development, the country has yet to initiate a clear set of policies that will guarantee the technological innovations and the improvements in industrial productivity that will be required to meet growing foreign economic competition, stimulate new economic growth, and solve many of our domestic economic problems in the future. Failure to design such strategies may indeed ensure that the American economy continues its relative industrial decline.

James Kurth and others have begun to advocate that, to facilitate a move out of present international economic morass, one alternative "would be for one country, most probably the United States, to undertake the development of new industrial sectors."[43] In fact, almost anyone who writes or talks about industrial policy these days comes around to the conclusion that it would be wonderful for governments to step in and speed up the process of technological innovation and create whole new sectors. There are limits, however, in what governments can and cannot do with their research and development programs in the matter of creating new technologies capable of generating new leading sector industries. Massive new infusions of R & D monies will not automatically result in the creation of new leading sectors; that is, unless these sectors are destined to emerge on their own.

Forecasters have been predicting the imminent arrival of new "leading" sectors — whether space travel, computers, telecommunications, or ocean development — for over a decade. The new leaders will emerge — they probably already are emerging, as the recent surge in microprocessor industries indicates. But they may not arrive in the near term, or at least they may not be ready to propel the rest of the economy forward soon. In the meantime, the American economy — and with it the prosperity of the American people — hangs in the balance. To concentrate vast resources on speeding the arrival of the "new wave" may be folly. Besides, it is not clear how desirable it is (probably extremely risky) for the state to try to target and, thereby, accelerate the emergence of new leading sectors. The recent history of the Concorde does not offer a great deal of confidence that governments can adequately predict the dynamics of the process that decides which advancing technologies become "spearheads."

Perhaps hardest of all the pressures to resist will be the growing view of industrial policy as primarily a program of "hitching the economy to the infinite."[44] Of course, government space programs, military weapons development contracts, and other large exploratory programs can serve an important function in helping to build up the structures of the most advanced technology industries. But for the middle and even near long-term, such programs are likely to drain more resources than they create; their value lies primarily in their demonstration effects, service of other goals, especially national security, and the potential very long-term payoffs once these advanced technologies are "routinized" into mass-production industries.

In addition, an industrial policy that concentrates primarily on facilitating development at the cutting edge of advanced technology and on protecting declining industries or helping transfer resources out of the "obsolete" technologies is an extremely risky one. Most existing European industrial strategies do just that — they accept the logic that "it is only possible to take effective action at the two extremities, the declining sectors and the 'peak' activities."[45] In essence, this is how Guy de Carmoy summarizes French industrial policy: "The French combine offensive strategy in favor of innovative industries in danger and defensive strategy in favor of declining industries."[46] Such a strategy should be imitated by the United States, but American industrial policy measures must not ignore the middle-term in deference to short-term exigencies and long-term projects.

Consequently, a successful industrial policy must channel resources toward less glamorous pursuits, such as encouraging a wide variety of civilian research and development programs and ever new juxtapositions of current technologies and industries. Such a strategy is one that is likely to buy substantial time by continuously creating "ripples" of new growth in the interregnum before the expected "tidal wave" of revolutionary technological advances. In the final analysis, it may be the ability of an American R & D policy to help generate continuous "ripples" that steers the American economy through its present difficulties and determines whether or not the United States can step back into a strong world leadership position by helping the rest of the world deal with its problems, too.

But, of even greater importance in the United States, industrial policies probably do not need to concentrate on large infusions of R & D funds designed to create a specific new leading sector. The whole history of American technological innovation belies any suspicion that progress on this front is dead without direction from the state. The danger is not only that the U.S. government will not do enough to create new leading sectors. It is also that the United States will prove itself less and less able to deal with the forces that fight change and seek to prevent those structural adjustments that would prepare the way for whatever new leading sectors that do emerge to prosper.

In short, a real dilemma facing the United States and the other advanced Western nations today is not so much that industrial policies are proving incapable of "dicovering" new leading sectors and more advanced technologies, it is that many industrial policies threaten to work against the natural dynamics that would bring forth such advances. Consequently, a key aspect of a rejuvenation strategy for the United States economy must be the design of industrial policies that complement heightened government and especially corporate R & D expenditures. These policies must facilitate the rise of new leading sectors by helping to head off the opponents of change and dealing with the economic and physical disruptions the new technologies create.

Most importantly, government interventions should concentrate on creating a flexible economy, in which the forces of production are mobile enough to move quickly into sectors whose time has come — in contrast to trying directly to create those sectors in a sort of "national champions" approach such as that of the French or British. In this sense, there is probably much that we can learn from the Japanese, who at present appear to be succeeding more than other major Western nations in devising industrial policy measures that encourage flexibility and innovation.[47]

Conclusions

It would be presumptuous to suggest that economic and technological forces will determine the course of domestic or international affairs. In the last analysis, passions rule the world. Nonetheless, to a degree perhaps unparalleled in the past, economic and technological considerations will shape the ways in which political interests and conflicts seek their expression and work themselves out. In a world where nuclear weaponry has inhibited the use of military power and where social and economic demands play an inordinate role in political life, the choice, success, or failure of a nation's technological strategy will influence in large measure its place in the international pecking order and its capacity to solve it domestic problems.

For the United States, the designation of a coherent industrial strategy to meet the future challenges of the international economy is of critical importance. A lot more is riding on America's ability to accomplish this challenge than the relative affluence of the American people. America's ability to look outward and exercise the leadership necessary to move the rest of the world toward peace and prosperity rests, more than anything else, on the institution of successful industrial policies that head off the decline or stagnation of the American industrial economy and the engine of the world economy.

In summary, the role of an industrial policy for the United States is to help ensure that the American economy remains vibrant and is strong enough to offset the costs of world leadership and help ease the pressures building up in the present political-economic international system. To meet these arduous demands, it is still true, as it was after the war, that:

> The greatest contribution we can make to economic order in the world is to maintain a strong, free, and stable economy at home. In some respects these are not entirely compatible objectives of policy. A free economy is a developing economy, liable to short-term interruptions of stability; but stability ought not to be regarded as synonymous with rigidity. Indeed, biology teaches us that flexibility within a narrow range is an essential element of organic stability.[48]

Although more and better programs to stimulate civilian R & D are essential, the task of ensuring flexibility and continuous innovation is not necessarily synonymous with initiating a massive government crash program to find a particular new leading sector.[49] New leading sectors will arrive in their own time and will help to fuel new spirals of growth and prosperity. Providing the United States uses its industrial policy to find new ways of recreating an economic environment that is accepting and encouraging of change, the American economy will remain in a position to take the fullest advantage of new waves of technological progress. In addition to maintaining mobility and flexibility, measures should be taken to ensure that big business and big government do not smother creativity. Finally, although strategies must be devised to meet welfare demands, they must be guided by the goal of softening the hardships brought by structural change, not turning back the change itself.

Notes

1. For more comprehensive descriptions of industrial policy and the range of government actions covered by the term, see William Diebold, Jr., *Industrial Policy as an International Issue* (New York: McGraw-Hill, 1979); Christopher T. Saunders, "Concentration and Specialization

in Western Countries, in Saunders (ed.), *Industrial Policies and Technology Transfers Between East and West* (Vienna and New York: Springer-Verlog, 1977); and Stephen J. Warnecke (ed.), *International Trade and Industrial Policies* (London: Macmillan, 1978). A related, but more narrowly conceived subject, is that of technology policy. The present paper incorporates parts of an earlier article by Gilpin on this subject entitled "Technology Strategies and National Purpose," *Science,* 31 July 1970, pp. 441–448.

2. OECD, *Policy Perspectives for International Trade and Economic Relations* (Paris: OECD, 1972), p. 34.

3. Charles-Albert Michalet, "France," in Raymond Vernon (ed.), *Big Business and the State: Changing Relations in Western Europe* (Cambridge, Mass.: Harvard University Press, 1974)

4. Göran Ohlin, "National Industrial Policies and International Trade," in C. Fred Bergsten (ed.), *Toward a New World Trade Policy: The Maidenhead Papers* (Lexington, Mass.: D.C. Heath, 1975), pp. 180–181.

5. Diebold, *op. cit.*

6. A good, though perhaps dated, study of lead times is *Technology in Retrospect and Critical Events in Science,* prepared and published for the National Science Foundation by the Illinois Institute of Technology Research Institute (Chicago,1968, 2 vols.).

7. See *Technology: Processes of Assessment and Choice* (Washington, D.C.: National Academy of Sciences, 1969).

8. One of the earliest works to survey the changing status of the field was H. G. Johnson, *Comparative Cost and Commercial Policy Theory for a Developing World Economy* (Stockholm: Almgvist & Wiksell, 1968).

9. For an early discussion of this phenomenon, see Charles A. Beard, *The Open Door at Home* (New York: Macmillan, 1935), pp. 210ff.

10. See Robert Gilpin, *U.S. Power and the Multinational Corporation* (New York: Basic Books, 1975).

11. See *The Aims and Instruments of Industrial Policy: A Comparative Study* (Paris: OECD, 1976).

12. Jacques Maisonrouge, quoted in "Neo-Mercantilism in the '80s: The World Wide Scramble to Shift Capital," *Business Week,* July 9, 1979, p. 51.

13. There are, however, some indications that Japan is introducing international acceptability as a more important criterion in its industrial policy decisions; see, Diebold, *op.cit.*

14. Lionel Stoleu, *L'imperatif Industriel* (Paris: Editions du Seuil, 1969).

15. OECD, *The Industrial Policies of the United States* (Paris: OECD, 1969), p. 5.

16. See, Robert Gilpin, *France in the Age of the Scientific State* (Priceton, N.J.: Princeton University Press, 1968).

17. See, for example, the discussions about France in Warnecke (ed.) op. cit., Steven J. Warnecke and Ezra N. Suleiman (eds.) *Industrial Policies in Western Europe* (New York: Praeger, 1975).

18. For a more detailed analysis of the strategy followed by Japan in the post World War II era, see P.B. Stone, *Japan Surges Ahead: Japan's Economic Rebirth* (London: Weidenfeld & Nicolson, 1969).

19. The literature on the Japanese miracle is rather extensive, but several studies are especially relevant: M.E. Dimock, *The Japanese Technocracy* (New York, Walker, 1968); R. Guillain, *Japon Troisieme Grand* (Paris, Editions du Seuil, 1960); H. Brochier, *Le Miracle Economique Japonais* (Paris, Calmann-Levy, 1965); OECD, *Reviews of National Science Policy-Japan* (Paris 1967): Terutomo Ozawa, *Japan's Technological Challenge to the West 1969-1979* — Motivation and Accomplishment (Cambridge, Mass.: MIT Press, 1974); Hugh Patrick and Henry Rosovsky (eds.), *Asia's New Giant — How the Japanese Economy Works* (Washington, D.C., The Brookings Institution 1976); Ezra F. Vogel, *Japan as Number One —*

Lessons for America (Cambridge, Mass., 1979) and Herman Kahn and Thomas Pepper, *The Japanese Challenge-The Success and Failure of Economic Success* (New York: Thomas Y. Crowell, 1979).

20. For an intelligent and balanced analysis of "Japan Incorporated," see, U.S. Department of Commerce, *Japan-the Government-Business Relationship,* (Washington, D.C., February 1972).

21. In addition, the United States has been pressuring the Japanese to make a greater contribution to the international pool of basic knowledge, which is available to all countries.

22. This section relies heavily on several authors associated with the Science Policy Research Unit of the University of Sussex, England: Christopher Freeman, Keith Pavitt, and W. B. Walker.

23. U.K. Government, *A Framework for Government Research and Development,* HMSO (Cmnd. 4814), 1971.

24. William Diebold, Jr., in Warnecke, *op. cit.,* p. 180.

25. George H. Kuster "Germany," in Vernon, *op. cit.,* pp. 73–74.

26. "Europe: The Investment Subsidies Keep Rolling," Business Week, August 9, 1979, pp. 37–38.

27. Zysman concludes:

The French state has attempted to maintain at least one national supplier of each politically important electronic technology. The dilemma has been that the protection and support required to produce specific products of interests to the state may in fact have weakened the firms that must be the long-term instruments of state policy. Without strong firms the development of new technologies cannot be assured in the future. Strong and healthy firms with competitive market strategies and efficient organizations must be the centerpiece of any successful government electronics policy, and financial mergers or joint ventures which increase the volume of capital will not prove a substitute for often painful change in the firm's strategies and organization. . . . [which raises] the need for companies to define and pursue competitive policies.

John Zysman, "French Electronics Policy: The Costs of Technological Independence," in Warnecke and Suleiman, *op. cit.* p. 245.

28. Hugh D. Menzies, "U.S. Companies in Unequal Combat," *Fortune,* April 9, 1979, p. 102.

29. Warnecke, *op. cit.* p. 12.

30. Jan Tumlir, "National Interest and International Order," *International Issues* No. 4 (London: Trade Policy Research Center, 1978), pp. 6–7.

31. Ohlin, *op. cit.* p. 180.

32. William Diebold Jr., *The U.S. and the Industrial World* (New York: Praeger, 1972), p. 252.

33. For an elaboration, see John H. Dunning, "United States Foreign Investment and the Technological Gap," in Charles P. Kindleberger and Andrew Shonfield (eds.) *North American and West European Economic Policies,* (London: Macmillan, 1971).

34. Thorstein Veblen, "On the Penalty of Taking the Lead," in *Imperial Germany and the Industrial Revolution,* excerpted in Max Lerner (ed.), *The Portable Veblen* (New York: Penguin Books, 1976), p. 375.

35. Jacques R. Houssiaux, "American Influence on Industrial Policy in Western Europe Since the Second World War," in Kindleberger and Shonfield (eds.), *op. cit.,* p. 353.

36. See *ibid.* for a more detailed analysis.

37. These strategies and their implications are elaborated in Gilpin, "Technological Strategies," *op. cit.*

38. For a dicussion of the interactions between states and societies in Western Europe, see Peter J. Katzenstein (ed.), "Between Power and Plenty: Foreign Economic Policies of Advanced Industrial States," International Organization, Vol. 31, No. 4, Autumn 1977.

39. For a recent analysis, see Ezra Vogel, *op. cit.*

40. The phrase is borrowed from Harold B. Malmgren, "Negotiation of Rules on Subsidies in a World of Economic Interventionism," in Warnecke, *op. cit.*

41. See also Ohlin, *op. cit.* p. 181

42. Indeed, a recent report by the Congressional Budget Office argues that there really is little the United States can do to slow the spread of adanced American technologies around the globe. The report is summarized in Hobart Rowen, "U.S. Finding Uneven Benefits in Multilateral Trade Proposal," *The Washington Post,* March 11, 1979, p. A32.

43. James R. Kurth, "The Political Consequences of the Product Cycle: Industrial History and Political Outcomes," *International Organization,* Vol. 33, No. 1, Winter 1979, p. 33.

44. The phrase is borrowed from a 1962 *Fortune* article that warned that even though space travel posed vast new economic horizons in the distant future, as well as rapid growth in certain government contract industries, it was no economic panacea: "Nor will the fabled practical benefits offset the cost of the program for a long time. Washington is teeming with lobbyists and other space partisans assiduously promoting the notion that space is the greatest sure-fire blue-sky investment ever, sure to pay off 1000 per cent immediately. . . . But it should also be obvious that no matter how abundantly it pays off, it will be a big and growing investment, and that, like all investments, it cannot be made until people first produce something to invest. To rise in the sky, the United States will have to keep its feet on the ground." Gilbert Burck, "Hitching the Economy to the Infinite," *Fortune,* June 1962, reprinted in John A. Larson (ed.), *The Regulated Businessman: Readings from Fortune* (New York: Holt, Rinehart and Winston, 1966), pp. 242–43, 255. Burck's warning is apropos today as well, as Washington is once now "teeming with lobbyists" promoting the notion that massive government investments in synthetic fuels and other energy sources will solve America's energy crisis overnight — and get the United States economy "on its feet again" in the process.

45. Houssiaux, *op. cit.,* p-. 352.

46. Guy de Coemoy, "Subsidy Policies in Britain, France and West Germany: An Overview," in Warnecke, *op. cit.*

47. Ezra Vogel claims: "Japan provides both crucial research and financial resources to spur economic and social development when private sources are inadequate. The United States should follow Japan's example by making an effort to distinguish those industries in our country that can be competitive on the world market and to support them through tax policy, monetary policy, and administrative cooperation. Further, it should make an effort to provide temporary cushions for those industries such as textiles, that cannot remain competitive. Americans must educate business and labor leaders to perceive and acknowledge these problems so they can, in turn, provide steadier predictable government policy." Ezra Vogel and Susan Chira, "U.S. Should Look to Japanese for Energy Answers," *Trenton Times,* Sunday, July 29, 1979, p. F-1.

48. J. B. Condliffe, "Foreward," in Henry Cholmers *World Trade Policies* (Berkeley: University of California Press, 1953), p. xii.

49. Simon S. Kuznets, *Secular Movements in Production and Prices — Their Bearing Upon Cyclical Fluctuations* (Boston and New York: Houghton-Mifflin, 1930).

4

Technology Policy and Policy Making in the United States

CHRISTOPHER WRIGHT

Introduction

The United States does have a technology policy. One can discern the broad policy outlines and note significant changes in its effectiveness over the past decades. The underlying purpose of the vast expansion of the nation's scientific and technical enterprise, stimulated and supported by government after World War II, has been to assure both an adequate defense and a leadership role in areas of high technology — a technology founded on scientific research and requiring highly trained talent. This is especially evident in such areas as aeronautics, astronautics, electronics, nuclear physics, and medicine.[1] Along the way, technical inventiveness, as such, has been rewarded, but no high priority has been given to the systematic development and exploitation of technological opportunities, wherever they may lead.

In fact, the vast proportion of the technological initiative of which the nation is capable neither depends upon nor is constrained by policies. Rather, most such initiatives are the expression of a "Yankee ingenuity," quite independent of government. To the extent this spirit now flags, the causes are to be found more in technological, social, and economic changes than in the characteristics of public policy. Present technology policy neither guides nor harmonizes with these forces. Nevertheless, technology policy has significant impacts over the long term, as the accomplishments of significant technological innovations require integration of the new technology into more and more complex social systems. The prospects for such innovations are strongly affected by whether or not this integration is hindered or facilitated by prevailing public policy.

Even though technology policy has made possible our awesome technological complex of defense systems, it has had strong traditional elements and has tended to be modified piecemeal. There are no effective processes for analyzing and altering policy in the light of changing circumstances and opportunities. Technology policy has tolerated, even encouraged, incoherent, wasteful, and counterproductive uses of our technological resources.

It was not created to develop and guide the nation's scientific and technical resources as a vital national enterprise. As a consequence, technology policy lags behind technological developments and solutions to major national problems. It does not guide them. Nowhere is this lack more evident than in the international context.

The nation seems incapable of exploiting its technological potential for specific national purposes. Laws have been enacted and new Congressional committees and Executive Branch offices created.[2] And yet, there has been remarkably little change in the process or substance of technology policy making since the early years after World War II. The remarkable changes in technological opportunities and in the nation's circumstances dramatize the declining utility of current technology policy.

A Federal policy making apparatus centering on technology as a national asset may not be preferable to the current state of affairs. It may not even be feasible, but it is apparent that, in the future, a broad range of international considerations will drive and shape U.S. technology policy, much as defense considerations have since World War II. The question is, will they give rise to a policy that will, in turn, provide sustained guidance for technological initiatives, or will this policy be essentially responsive, constrained by the focus and format of past policy?

U.S. technology policy now reflects predominantly domestic considerations more than the international leadership role of the nation. Traditional dichotomies between civilian and defense technologies and between domestic and foreign affairs tend to be preserved, even though the inherent nature of many new technologies are inconsistent with such divisions. While the traditional disposition to welcome and reward the technical innovations of individuals persists,[3] those innovations that might be of greatest social and economic importance but require much more collaboration and coordination are increasingly constrained by an accumulation of laws, regulations, and standards which effectively favor established techniques, equipment, work places, or modes of technological innovation.

Present technology policy exhibits a reasonable, responsive, yet essentially passive character which, nonetheless, has increasingly serious international consequences. It is easier to exploit new technologies for sending astronauts to the moon and back than to improve transportation services in a metropolis. Technologies are developed and exploited more because they can be accommodated within existing institutional and organizational arrangements than because scientific findings create technological advances and opportunities.

Moreover, when technical solutions or "technological fixes" to social problems are advanced, there is little capacity for taking account of their implications for broader social developments here or abroad. As yet, technology assessment is little more than an idea and a felt need, despite the creation of such units as the Congressional Office of Technology Assessment. Such efforts are overshadowed by the preconceptions and political voices of those who see themselves as most affected, one way or another, by a new technology.[4]

Full exploitation of the nation's technological potential must await establishment of organizations whose purposes are to achieve social ends with the help of the scientific and technical enterprise. To be viable, such institutions will have to have the political standing — the clout — to overcome resistance to such uses of technology.

Origins of Postwar Technology Policy

The mainstream of America's scientific and technical enterprise is fed by many springs. In the past it has flowed freely, often cutting new courses into the economy. That situation is changing. In place of spontaneity, today we see highly organized research designed to uncover the potential of a specific technology or line of scientific inquiry, such as nuclear fusion, agencies created specifically to advance some line of technological innovation such as the Atomic Energy Commission, and others designed to regulate the expression of technology such as the Environmental Protection Agency.

The first expression of post-World War II technology policy can be found in the report, *Science: the Endless Frontier,*[5] prepared in 1945 by Vannevar Bush, then Director of the wartime Office of Scientific Research and Development and a principal adviser to President Roosevelt. The Bush report recognized that, henceforth, the Federal Government would have to take an active interest in the promotion of scientific research and personnel, the sources of technologies shown to have revolutionary implications for defense. At the same time, Bush assumed the importance of private initiatives. He did not encourage a central management role for the Federal government in the scientific and technical enterprise. In fact, he proceeded to dismantle his Office of Scientific Research and Development.

The Bush report looked for unexpected, broad-ranging technological benefits from the promotion of the natural sciences. Subsequently, the nation's commitment to Federal support of research and training has been expressed in the creation of granting agencies and major government laboratories, and in the production of many reports. Funding programs include those of the mission agencies and the National Science Foundation. National Laboratories, including the National Institutes of Health, have been established to meet needs that could not be met otherwise. Reports initiated by the Congress, the President, and private organizations, have been prepared with the aim of anticipating and guiding the uses of science and technology to meet the nation's needs and objectives.[6]

Explicit policy statements are contained in the *National Science and Technology Policy, Organization, and Priorities Act of 1976* (P.L. 94–282) and in other recent legislation. Much of the language of such acts calls on the Executive to engage in policy formulation and implementation, an aim yet to be achieved. Machinery for addressing policy issues associated with science and technology has taken the form of Congressional oversight committees for science and technology, the Office of Technology Assessment, and the Office of Science and Technology Policy in the Executive Office of the President. These pieces of machinery do not yet function as their creators expected.[7]

Despite formal recognition of the need to address technology policy issues, a vast array of federal programs, expenditure patterns, laws, and regulations composing the main body of U.S. technology policy are beyond the purview of any one entity. The Defense Department and NASA head the list of agencies with procurement programs that go a long way toward determining, not just implementing, technology policy. The NSF, the NIH, the Department of Energy and the other mission agencies support research and development programs designed to accomplish the missions of the particular organizations, but not necessarily to further an overall national technology policy.[8]

Regulations designed to protect health, enhance safety, or protect the environment have tended to inhibit technological innovations in some areas and to encourage them in others. Sometimes, as with the introduction of new drugs, the regulations are intended to slow the rate of innovations so as to assure their efficacy.[9] Occasionally, as in the case of auto emission standards, regulations seek to force the development and introduction of a new technology on a predetermined schedule. Different facets of technology policy are reflected in such regulations, but there is no way to assure that the priorities they express are compatible and especially effective in terms of national priorities for exploiting technological opportunities.

Economic incentives and regulations also play a fundamental role in shaping technology policy. Patent and tax laws are increasingly ill-suited to encouraging the long-term strategies now required to bring about major technological innovations. Economic regulations designed to maintain business competition are not well adapted to the new kinds of industrial consolidation, differentiation, and competition implicit in technological innovations, such as those which exploit electronics.[10]

Laws protecting property rights and assuring due process usually favor the status quo and hence inhibit or preclude technologically induced changes. The only exceptions are where national security or public health and safety mandate the changes.[11] The Chrysler Corporation, adversely affected by changing automotive technologies, may be saved by specific governmental initiatives designed to support endangered industries.

Hand in hand with these impacts of economic policy on technology policy go financial and legal judgments. The influence of lawyers, financiers, and accountants, in contrast to that of engineers and scientists, is great. Lawyers and financiers occupy leadership positions in industrial corporations. Legal and financial considerations tend to shape corporate strategies more than do the opportunities for making significant technological innovations.[12]

Against this background, legislation addressing technology policy has thus far been more an expression of Congressional intentions than of significant operations. It has not addressed the sorts of factors shaping technology policy mentioned above. It has not produced more policy consistency and coherence. A coherent policy would inevitably cut across many areas of divided legislative responsibility in the Congress, as well as agency responsibilities in the Executive Branch, and thus be extraordinarily difficult to achieve. One modest approach to overcoming this problem that seems especially well-suited to the ways of Congress is the formation of ad hoc groups of likeminded members able to play complementary roles in various legislative areas. In the House of Representatives there is now an informal Industrial Innovation Task Force and a publication dubbed *Congressional News Notes on Innovation and Production,* sponsored by the House Subcommittee on Science Research and Technology.

One legislative proposal that could alter and implement technology policy in a major way is a bill introduced in 1980 by Congressman George E. Brown of California that would create a National Technology Foundation with both funding and operational capabilities. Reactions to this proposal have heightened awareness of the potential importance of public technologies that are neither the concern of private companies with proprietary interests, nor of the established scientific and engineering disciplines. Attention is being drawn to the close interactions among the sciences, engineering, and new technologies, and also to the growing institutional differences between scientists and engineers.

Otherwise, there have been few efforts in recent years even to identify the disparate elements of technology policy with the aim of eventually enhancing its effectiveness.

One such effort was the provision of the 1976 act calling for a President's Committee on Science and Technology to conduct a sweeping survey and analysis of many aspects of science and technology policy making and their implications. The act mandated periodic presidential science and technology reports on developments and trends, and a separate report identifying major problems and opportunities associated with science or technology likely to emerge within a five-year time frame. The Committee's task was not accomplished and the periodic reports have not, as yet, met the expectations of the Congress.[13]

The Limitations of Current Technology Policy

Given the present state of technology policy, it is not surprising that there is little regard for efforts to set specific national priorities.

There seemed no need to establish priorities while the nation could reasonably expect to be at the forefront of every military and industrial technology and field of science. Now, the United States is demonstrably no longer the across-the-board leader. From a technological point of view, the United States is falling behind in such diverse areas as steel and shipbuilding, chemical manufacturing, automobiles, consumer electronics, and in other still more sophisticated technologies.

No nation can maintain across-the-board technological leadership. Few nations compete in such areas as aircraft and electronic computers. Even fewer strive to maintain military superpower status. Some advanced technologies and areas of scientific research invite collaborative efforts among companies and nations. Others, such as semiconductor technology, tempt nations to exploit the new technologies in areas which, with earlier technologies, were the prerogatives of those with the greatest access to capital, markets, or essential resources.[14]

Despite such changes in technological circumstances, there is no systematic approach to involving the nation's scientific and technical enterprise in its international relations. There is not even an effort to identify the mechanisms for developing such an approach.

Granted the subordinate role of technology policy in national policy making, there is, nevertheless, an important sense in which technology policy, as well as specific technological advances, have affected the nation's international position. Technology policy has tended to reinforce the nation's emphasis on innovative military defenses over innovative civilian technologies.[15] It has discouraged substantial responses to international technological competitiveness,[16] and it has generated only modest efforts to mobilize science and technology for the benefit of developing areas.[17]

There are underlying reasons why U.S. technology policy has not given more direction to the impact of technological changes on international relations. Increased institutionalization of the scientific enterprise is one reason. It tends to compartmentalize scientific thinking and discourage its unorthodox use. Another is that dramatic changes brought about by scientific advances generate doubts about managing those changes and cautions to limit their scope.

Yet another reason for the passive state of U.S. technology policy has been a certain euphoria about the fruits of science and science-based technology. The outcome of World War II suggested that the United States owed its newly acquired world status in large measure to its technology-based industrial enterprises, to its reservoir of natural resources, to a relatively well-educated populous, and to a large and open community of scientifically and technically trained persons.

Science: The Endless Frontier, by Vannevar Bush, was one of the initial efforts to translate that insight into policy. Vannevar Bush was an engineer and inventor, as well as a science administrator. He recognized the differences between science and engineering and appreciated the importance of creative interactions between them. He also recognized that the most vital contributions of science and technology to any particular social mission may come from unexpected quarters and hence should be advanced on that basis.

The title of his report emphasizes this potential of science, but the metaphor of the frontier overlooks problems for established institutions of making full use of science and technology as a national resource. "The endless frontier" suggested exploiting new riches without disturbing the old lands. Modern science was accepted on this basis. The dominant belief in science has been that natural processes are governed by essentially simple physical principles, and, hence, should be viewed in terms of one-to-one relationships between causes and effects. Physics provided the preferred model both for science and for its technological applications.[18]

Now, as the potential for wide-spread consequences of new technologies is recognized, social and natural processes are viewed as having many subtle interactions. Mental models based on biological and ecological systems are supplanting those of physics and chemistry. Consider the concepts and vocabulary associated with computers and computer programming — the importance of "feed-back" systems, for example.

As a nation, we have resisted the implicit reordering concealed by the manifest destiny of "the endless frontier." But over time, the value-free aspects of science have become overshadowed by the social and technological changes wrought by the opening of new scientific frontiers.

Our technology policy does not yet embrace the concept that exploitation of new technologies is likely to have unexpected consequences for interests and values as it effectively meets specific needs, strengthens some social interests, and reinforces certain values. The notion that such impacts of technology can be anticipated and prepared for has emerged slowly.[19] In part, this is because we still lack conceptual tools to deal with them, and in part because of differences both in perceptions about technical causes and social consequences and also in beliefs as to the need to assess them.

The direction and priorities provided by policy are promoted by those who have a particular interest in the technical opportunities and initiatives.[20] Consequently, policy is viewed in terms of existing institutions, laws, and regulations, rather than as principles intended to guide the interactions of initiatives and social purposes.

There are two ways to address the inherent complexity of the relationships between new technologies and their social impacts. One is to develop comprehensive technical plans and implement them through a central authority.[21] The other is to use whatever government authority is available to devise policy statements that both acknowledge inherent scientific and technical imperatives and are widely enough understood so that persons with diverse responsibilities, each acting independently, may proceed in ways that are in harmony with policy. The first approach tends to extend the role of technical expertise beyond its competence and discounts the views of affected persons. In most situations it lacks political legitimacy. The second approach is more acceptable, but it depends on a skill in policy formulation that remains underdeveloped, especially with respect to technology policy.

Technology policy is not yet providing effective guidance for how to mediate the interactions between established policy concerns and major new technological opportunities. It has provided ways to readjust the rates of technological innovation in selected areas, to increase domestic and international collaboration in some areas and competition in others, and to develop incentives for technological endeavors in unconventional situations. This is particularly evident in regard to international relations, as reflected in the technological initiatives related to such areas as national security, Third World development, natural resources, and industrial competitiveness. These areas of international challenge also illustrate the range of opportunities and the serious limitations of current technology policy as a key instrument of U.S. foreign policy.

Technology Policy and National Security

Since World War II, the process of weapons systems development and procurement has been supported by a vastly expanded scientific and technical enterprise. That enterprise has created governmental machinery for making and implementing technological plans. While technology policy making has been shaped by national security concerns, its effectiveness has been due to the ways in which it reinforces established industrial, educational, and research institutions. Now, major changes are occurring. Implementation of national security policy may upset the plans of industrial companies or of the scientific community, as when technology transfers and scientific exchanges with the Soviet Union are abruptly terminated. Implementation efforts can also be insensitive to technical analyses which appear to question established approaches to national security objectives.

Prior to World War II, the Army and the Navy relied on government arsenals and shipyards for technological innovations. After the War, the Army co-opted German rocket capabilities, the Office of Naval Research took the lead in supporting university-based research, and the newly autonomous Air Force initiated a policy of industrial collaboration for development and procurement. The Atomic Energy Commission supported physics research and education in part to assure the rapid development of nuclear weapons. Through its Industrial Research and Development program, the Department of Defense encourages private companies to develop new technologies.

By means of this pattern of mission agency support, technology policy has encouraged a succession of technological innovations related to a national security policy centering on strategic nuclear deterrence.[22] The most modern scientific and technical resources are sustained in large measure because of their assumed importance for national security. And yet, technology policy has also tended to limit the ways in which those resources can be mobilized to address changing international conditions.

Drawing on the contributions of scientists in World War II, the technology policy enunciated in the Bush Report and elsewhere has encouraged a continuous flow of advanced science and technology into the defense area.[23] The traditional conservatism of the uniformed services has been transformed into an enthusiasm for technological innovations that serve the needs of the armed services. With the synergistic interactions between the Federal government as procurer of new technologies and private companies as promoters thereof, one new military technology has proceeded another at a remarkable rate. In many areas, technologies have been pushed to the brink of what is feasible, sometimes with scant regard for their actual utility or cost-effectiveness.[24]

In the absence of a technology policy that would couple technologies with national goals, it is questionable whether weapons technology advances have advanced national defense and security as much as they have advanced the role of the agencies involved. Some scientists and technologists-turned-weaponeers were among the first informed persons to point out that successive advances in nuclear weapons technology were tending to make the delicate balance of mutual deterrence less stable — and thus reducing the security of the nation. It is arguable that nuclear arms control may be a more viable political (and technical) objective. Similarly, the assumption that a "nuclear umbrella" can limit, if not prevent, conventional warfare is being questioned by those who have watched its effects. Even the benefits of extreme sophistication of equipment are now being questioned in the light of experience that shows human capabilities being outpaced and overtaxed by new weapons and new command and control systems.[25]

As the technological capabilities of other nations advance relative to ours, the United States no longer has the same degree of control over international technology transfers. The national security policy issues surrounding such transfers are complex and unresolved. Indeed, the issues change as rapidly as the technologies themselves. Often, efforts to deny technologies to unfriendly countries may have more long-term adverse consequences for us than for them. The problem is that current modes of policy analysis are not likely to reveal the costs and benefits in these terms.

Vannevar Bush and his colleagues anticipated the need for lines of inquiry about technological opportunities. Subsequently, technology policy and policy makers have not especially encouraged them. Instead, economic policy makers have taken the initiative by encouraging cost/benefit and cost/effectiveness analyses, as was done in the Defense Department under Secretary MacNamara, sometimes to the dismay of those who cast themselves in the role of advocates of unfettered technological innovation.[26]

In part, as a consequence of a preoccupation with old types of national security technologies and missions, the international leadership role of the United States is diminishing. The nation is perceived not only as striving to persuade other nations to accept the terms of military technological competition as we see them, but also as blind to the technological opportunities so important both in international competition and in our search for essential resources.

Our technology policy does not encourage ongoing reassessment of the ways in which technology might relate to and anticipate the changing meaning of national security and the national interest abroad. Instead, it encourages the Defense Department and related mission agencies to acquire technologies that enhance their organizations, but do not necessarily accomplish the missions for which they were established. Scientific findings that might reassert or redefine their missions, and hence the associated technologies, are not encouraged. The prevailing assumption is that the technical means of most interest to a mission agency are those that it knows how to manage. Hence, arms technologies are of interest to the DOD (Department of Defense), but the technical studies that might better be described as peace research, leading to peace keeping systems, are not.

The national security establishment's close coupling with the scientific and technical enterprise accounts for the tremendous postwar enlargement of the latter. But there has been a price. The attention of too many exceptionally talented and highly trained persons tends to be absorbed by predetermined tasks. The preordained defense preoccupation contributes to the difficulty the United States has had in addressing scientific and technical resources to nonmilitary areas: assisting in the development of Third World nations, assuring the availability of vital resources, and competing with other industrialized nations.

Third World Development

One of the major developments since World War II has been the emergence of developing countries as nations with strong voices, obvious economic and social needs, and little military power. It has been deemed in our national — and security — interest to assist development. The long-term solutions to economic and social development will have to draw on the full and appropriate use of all that science and technology have to offer.

The nation's efforts to meet this challenge are limited by a technology policy that places responsibility for initiatives on private corporations, on mission agencies, on individual inventers or entrepreneurs, and on the established technical disciplines and professions. And yet, the solutions to the problems of Third World development are not likely to match the capabilities of any of these parties acting independently.

Despite periodic efforts to engage science and technology in this effort, the results thus far have been disappointing. Our technology policy is by no means the only cause, but it has contributed. The valid policy aim of assuring technical competence, continuity, and unambiguous progress tend to reinforce established fields of research and development — and discourage new ones more appropriate to the needs of development.

The fate of the proposed Institute for Scientific and Technical Cooperation illustrates the limitations of our technology policy. ISTC was advanced by the Carter Administration in 1978, but official United States interest in the suggestion that the nation could and should harness science and technology for the benefit of developing countries started with the 1956 United Nations Atoms for Peace Conference, if not before. In 1963, a more comprehensive approach to the subject was embodied in another UN Conference. And in 1978, the United States once again participated most actively in the UNCSTD Conference on the same subject. At that Conference, the U.S. Delegation announced the creation of ISTC as the centerpiece of its plan to bring science and technology to bear on development problems in a more effective way. Subsequent Congressional and Executive Branch reservations, however, led to the early demise of the proposal.

The proposal for a government-supported unit detached from regular mission agencies, but devoted to the uses of science and technology for development, has been advanced by concerned scientists and engineers for a decade or more.[27] They see it as a way to get involved in government programs without having their freedom of inquiry excessively constrained. In its management structure and scale of operations, ISTC was a compromise that satisfied neither government agencies, nor the scientific and technical community. The ISTC would have had to have substantial independence from operating agencies and yet serve as a gadfly with respect to them. It would have had to give voice to the ideas of scientists and engineers and yet be responsive to the perceived needs of developing nations. For the scientific and technical community, ISTC might also have been a disturbing gadfly.

Properly directed, ISTC could have met the need for new kinds of research and for professionals capable of addressing many-faceted development problems. Our universities and private companies are not equipped to conduct the most essential research or training and can not become so except as part of a long-term strategy. Present technology policy takes little account of the kinds of reliable long-term incentives required for this purpose.

A shortsightedness is built into U.S. technology policy's reliance on grant and contract modes of support for scientific and technical endeavor. Since World War II, the declining purchasing power of R & D budgets has combined with increasingly stringent forms of accountability to narrow the vision of technology policy. At the same time, such problems as Third-World development point to the need to legitimize long-term strategies for developing fields of science and technology that do not necessarily relate to current or traditional needs.

The "Green Revolution" in Third World food production through new agricultural technologies was possible because of the applied science work initiated for this purpose more than 20 years beforehand. The initial scale was modest, but there was no doubt that the effort would require many years of work, and it did. Technology policy today does not take account of such lessons.

Natural Resources

Since World War II, the nation's natural resources position has changed radically, from one of abundance to one of scarcity. The few early warnings[28] from the scientific community were not translated into technology policy. Until the oil embargo of 1973, it was easier to assume that foreign sources of oil and minerals would always be available. Technological resources tended to be devoted to increasing production efficiency, rather than to exploration for alternative sources or to development of substitute materials and forms of energy. The fact that such activities usually require long lead times increases the difficulty of reconciling the priorities of the nation and those of resources companies.

Likewise, government departments concerned with defense, finance, trade and commerce, public lands, environmental protection, and the earth sciences have differing positions on resources issues. This is evident in the *Global 2000 Report*[29] prepared at the request of President Carter, and in the abortive effort to conduct the Nonfuel Minerals Policy Review, which President Carter also ordered.[30]

The international position of the United States is affected in part by a technology policy that has not encouraged early warning or orderly accommodation to the technological implications of reduced availability of mineral resources, of increasingly precarious sources of supply, and of intensified competition for what remains. These concerns prompted the Congress to enact the *National Materials and Minerals Policy, Research and Development Act of 1980* (P.L.96–479) The Act is indicative both of the belief that our technology policy does not provide adequate guidance in these and similar matters, and of the increased sense of urgency about developing a technology policy appropriate to these times.

The Act requires that the President and various Cabinet officers plan for a technology policy that will lead to enhanced research, improved technologies, an adequate supply of technically trained personnel, and better resources information and early warning on supply problems, all in order to "ensure adequate, stable and economical materials supplies essential to national security, economic well-being, and industrial production."[31] The Act also calls on the President to assess Federal policies that adversely or positively affect all stages of the materials cycle and to assess opportunities for international cooperation for these purposes.

This is but the latest effort by the Congress to get the Executive Branch to develop a technology policy that is responsive to many fundamental concerns about the technical, economic, and political relations between national security and the requirements of defense systems, on the one hand, and industries, foreign policy, and access to strategic minerals, on the other.[32]

Industrial Competitiveness

Another obvious development since World War II has been the nation's loss of international leadership and competitiveness in some key industrial sectors.[33] The critical problems are thought to be declines in productivity and in industrial innovation. The most severely affected industries seem to lack the capacity for technological innovation. In many industries, there may be a "not-invented-here" syndrome, which causes companies and whole industries to disregard evidence of innovations from abroad. Nevertheless, as a matter of technology policy, it is assumed, as with agencies of government, that technological innovations can and should emerge from among the competitors in the industry that needs them, or at least be recognized and adopted by them.

Now, issues of technical collaboration among competing companies and with the government are coming to the fore,[34] as are issues of international technology transfer. Many inconsistent aspects of policy in both areas are at least being identified. As noted above, technological advances often lead to the amalgamation of sensitive military and civilian applications, to multinational businesses, and to the overlapping of heretofore distinct industries operating under different laws, economic rules, or government regulations.

The review of innovation policy ordered by President Carter was disappointing to the Congress because it failed to address these issue areas and such underlying impediments to innovation as the tax structure or assessed the policy implications of basic technological trends.

Techniques for disseminating scientific knowledge, technical know-how, and the movement of goods and services on a world-wide basis have altered the nature of trade and the meaning or dependence and interdependence in ways that now seem unmanageable in terms of current foreign or domestic policy. In the automobile industry, international competition is replacing domestic competition. New manufactures based on semiconductors are suitable for world-wide markets and may even require them. Resource-rich countries may process their natural resources, thus adding value to their value and competing with the importing industrialized nations. As military allies, nations have compelling reasons for sharing military technologies, and yet these same technologies could be the basis for civilian product areas in which they are in competition.[35]

Evidently, our confidence in free enterprise and competition does not carry over to changes in industrial competition wrought by major technological innovations here or abroad. And yet, we are surrounded by such changes with little guidance from technology policy. Separate industries that were governed by quite different public policies, such as telecommunications and data processing, are overlapping to an increasing extent. New industries, like the one now building around recombinant DNA technology are raising unprecedented regulatory, legal, and institutional questions about proprietary techniques and knowledge, and the patenting of living, reproducing organisms. For lack of alternative mechanisms, such issues are left to the courts, political processes, private enterprises, or government agencies to resolve in their own ways, but without much guidance as to how their actions affect those of others, and vice versa.

The Congress is aware of the need for a more coherent technology policy in order to give greater coherence to our foreign relations and to the maintenance of our domestic economy. Through the Acts cited above, it has also sought to assure that the nation's scientific and technical capabilities will be utilized consistently in relation to such aspects of international relations as national security, Third World development, resources dependencies and independence, and industrial competitiveness. It has also sought to have the Executive Branch to assess how different government programs and policies relate science and technology to American diplomacy.

Title V of the *Foreign Relations Authorization Act, Fiscal 1979* (P.L. 95-426) entitled "Science, Technology, and American Diplomacy," declares a policy of applying science and technology to problems of foreign policy and assigns related coordinating and oversight responsibilities to the Secretary of State. As with the other Acts, the language of this Act is promising. In practice, it is proving difficult to implement, even to the point of assembling pertinent information and developing a framework for comprehending the scope of technology policy in this context and the issues that it must address if the declaration of the Congress is to be fruitful.

Part of the problem of implementation rests with the Executive and the incapacity of any one department to adopt a perspective that is not centered on that department. The much larger problem is that the Congress simultaneously enacts legislation and approves appropriations without regard for the real possibility that in so doing they are contravening declarations of general policy that they have approved elsewhere.

Conclusion

Neither clear planning, nor a clear technology policy have guided the United States approach to national security, to relations with the Third World, to natural resources, nor to industry. It is not too late to remedy that situation.

By now it is clear that significant changes in technology policy are going to be extremely difficult to accomplish because present policy reflects established institutions in the scientific and technical communities, in industry, and in the Federal government. Basic changes in policy will require new concepts that will help everyone understand what roles technology might play in international affairs, what policy choices there are, and how they can be implemented. It may also require changes in the relationships among these institutions in the direction of voluntary coordination in terms of policy guidelines, with each entity performing the functions it can do best.

In each of the four areas discussed above it is clear that an essential interdependence has evolved among the private sector, the scientific and technical community, and the Federal government. No one part can adhere to a constructive technology policy or implement policy objectives without understanding how its activities might complement and reinforce those of the others.

Over the past decades, the technology policy of the United States has lagged further and further behind that of other nations, as technologies have changed and their role in shaping international affairs has increased. Belatedly, the Congress has sought to set in motion processes for devising a policy that will not only be commensurate with this age of science and technology, but will provide constructive guidance for exploiting technologies to enhance the nation's position.

Notes

1. Total federal obligations for research and development went from $15.6 billion in 1969 to approximately $28 billion in 1979, of which the R & D expenditures associated with the national defense, space, and health functions alone went from 84 percent to 72 percent of the totals. (From Table 2-16, *Science Indicators* 1978, National Science Board, 1979.)

From 1960 to 1977, industrial R & D expenditures in the areas of aircraft, electronics and communication, and drugs and medicine went from $7.4 billion to $8.4 billion in constant (1972) dollars, which amounted to 48 percent and 40 percent respectively of the total industrial expenditures on R & D. Over the same period of time, R & D expenditures in the areas of industrial and other chemicals, petroleum refining and extraction, and motor vehicles went from 19 percent to 22 percent of total industrial R & D expenditures. (From Table 4-2, *Science Indicators,* op. cit.)

In 1961 and 1976, the percentages of total governmental funds for research and development associated with national defense, space, and health in 1961 and 1976 in the following industrialized nations were: United States — 88 percent and 74 percent; United Kingdom — 60 percent and 51 percent; France — 46 percent and 39 percent; Federal Republic of Germany — 22 percent and 20 percent; and Japan — 5 percent and 10 percent of the totals. (From Table 1-5, *Science Indicators,* op. cit.)

2. Following the Soviet launching of Sputnik I in 1957, the House of Representatives created a Committee on Science and Astronautics, which subsequently became the Committee on Science and Technology. The Senate created a Committee on Aeronautical and Space Sciences. In 1977 that Committee was merged with the Commerce Committee to become the Committee on Commerce, Science and Transportation, with a Subcommittee on Science, Technology and Space.

By Executive Order, President Eisenhower created the Office of Science and Technology (OST), giving it science and technology responsibilities that had lain dormant in the National Science Foundation.

The National Science and Technology Policy, Organization and Priorities Act of 1976 (P.L. 94-282) includes declarations of priorities and policy and also mandates an Office of Science and Technology Policy (OSTP) in the Executive Office of the President. An authoritative account of the activities of OSTP from 1977 to 1980 is contained in a two-part article by Frank Press, Science Adviser to President Carter and Director of the Office for that period: "Science and Technology in the White House, 1977 to 1980" (*Science,* Vol. 211, 9 January 1981, pp. 139–145 and 16 January, 1981, pp. 249–256.)

3. Article I, Section 8, paragraph 8 of the United States Constitution states that the Congress shall have power "to promote the progress of science and useful arts, by securing for limited times to authors and inventors the exclusive right to their respective writings and discoveries."

4. The Technology Assessment Act of 1972 (P.L. 92–484) created the Office of Technology Assessment in order to assist the Congress in its anticipation, understanding, and consideration of the consequences of technological applications in determining public policy (Sec. 2). Its basic function is "to provide early indications of the probable beneficial and adverse impacts of the ap-

plications of technology and to develop coordinate information which may assist the Congress." (Sec. 3) In practice, considerations of impacts have tended to be recast as assessments of the feasibility of applying a technology. The interests of Congressional committees and the skills of analysts tend to favor the latter interpretation over the former.

5. *Science: The Endless Frontier,* A Report to the President on a Program for Postwar Scientific Research by Vannevar Bush, Director of the Office of Scientific Research and Development, July 1945. 192 pages. (Reprinted May 1980 by the National Science Foundation.)

6. Starting with President Hoover's Commission on Recent Social Trends and continuing with President Roosevelt's National Research Council, President Eisenhower's Commission on Goals for Americans, and President Nixon's National Goals Council, there has been a sequence of Presidentially commissioned reports on the future of the nation. Inevitably, the vital importance of science and technology to the nation is stated, but with few specifics as to the key aspects of an effective technology policy. Most recently the report of the President's Commission for a National Agenda for the Eighties (1981) states that, "The health of science and technology and the health of the economy are reciprocally linked: progress in the former generates growth in the latter, which in turn provides further resources for research and development." (p. 52). Nevertheless, the report contains few related policy recommendations, noting only that science and technology require certain kinds of support and management if they are to meet legitimate human needs. (p. 155). The report makes, but does not pursue, the observation that, "Reliance on the private sector to make adequate investments and to select technological options that are viewed as acceptable by the public has to be weighed against the demands that the government do whatever it can to solve perceived problems." (p. 156).

7. The report on a National Agenda, referred to in the previous note, also observes that it is not sufficient to codify policy, as is done in the 1976 Act (See Note 2). By calling both for an active commitment to such a policy through the allocation of money and personnel, and also for a stronger OTA, the Commission is tacitly admitting that statutory technology policy is not effectively implemented (pp. 159–160).

8. Although the NSF is the most general purpose supporter of research, it is primarily committed to the support of the research efforts in universities and, hence, has great difficulty furthering those kinds of basic and applied research or technological development that cut across disciplines or are otherwise beyond the scope of our academic institutions.

9. It is easier to invent and secure a proprietary position for a new drug and market it successfully than to assure its efficacy. Accordingly, the Federal government attempts to assure efficacy through regulation and licensing, rather than through support and control of research. The drugs and medicines sector of industry has the lowest proportion of federally funded R & D to company funded research and by far the highest ratio of basic research to net sales (Science Indicators 4–10).

10. Market competition in component parts of complex communications and computer systems depends on the acceptance of standard interfaces among the components. But, agreements for this purpose may also tend to freeze the overall course of technological development in that area. The setting of uniform standards may also have inherent advantages for some competitors over others, and, hence, have some elements of an illegal trust or cartel arrangement.

11. Even so, technological initiatives of importance to national security, such as the installation of missile or electronic systems that affect private property, have delayed or changed for that reason. And, despite the public health hazards of cigarette smoking in public places, the implementation of that imperative has been impeded in many ways.

12. In contrast to the United States, Japanese chief executive officers in industry and government are more likely to have been trained as engineers than as lawyers.

There are fewer than 12,000 lawyers in Japan as compared with about 500,000 in the United States. The relative absence of litigation in Japan and the preference for conciliation is noted by

Philip H. Abelson in an editorial, "East is East and West is West" (*Science,* February 6, 1981) and by Adam Meyerson, "Why There Are So Few Lawyers in Japan" (*Wall Street Journal,* February 2, 1981). Meyerson also notes that an American lawyer, Helen I. Bendix, has pointed out that, "American lawyers often play pivotal non-legal roles. With their verbal, analytical and deal-making skills, lawyers in U.S. companies often advise on business decisions and participate in complex negotiations with investment bankers, accountants and other outside specialists."

13. *The Five-Year Outlook: Problems, Opportunities and Constraints in Science and Technology,* National Science Foundation 1980, 2 volumes. The report describes the directions science and technology are taking, as viewed by the National Academy of Sciences, and the uses that the various governmental departments expect to make of science and technology. It does not, however, attempt to advise the Congress concerning the related implications for its future legislative program responsibilities, nor, as a statement prepared by NSF and its Director notes, does it carry the imprimatur of the President and his Science Adviser.

14. See *Science Policies of Industrial Nations* (edited by T. Dixon Long and Christopher Wright), Praeger Publications, New York 1975.

15. See Note 1 and *Science Indicators,* Chapter 1, for a discussion of corporate science and technology. See also Table 1-5 for comparisons of proportions of government R & D expenditures in national defense, and page 13 and Table 1-8 for the amounts of nongovernmental, as well as governmental, performance of research and development.

16. *Science Indicators,* Table 1-15 shows that among six leading industrialized nations, the United States had the least growth in productivity in manufacturing industry for 1960 to 1977.

17. The promise of such international cooperation has been reflected in many bilateral and multilateral agreements, as noted by Frank Press (see Note 2), but tangible accomplishments are elusive. Most recently, President Carter sought to create a modest but potentialy innovative Institute for Scientific and Technological Cooperation (ISTC), which failed to gain sufficient Congressional support.

18. Governmental organizations responsible for the successful development of nuclear weapons and the promotion of nuclear power as a source of energy draw heavily on the physics community for their leadership, but they have been slow to anticipate the full social, strategic, economic, political, and environmental implications of their technological initiatives.

19. The creation of the Office of Technology Assessment by the Congress expressed a well-intentioned desire for such a capability. Unfortunately, performance of the function depends on kinds of professional competence that will have to be instilled through our universities. As yet, the structure of higher education does not encourage such a development and there are insufficient incentives to change the system.

20. The plan, a decade ago, for the Federal Government to underwrite development of a supersonic transport plane (SST) was a prime example of a technological initiative pursued as an extension of existing interests and policy guidelines for the promotion of civilian air transport, with only a belated counter interest in discovering the inherent limits of acceptability for this technology in terms of social purposes.

21. The case for central planning is enhanced by the success, in limited contexts, of the NASA Apollo Program and the Naval Nuclear Submarine Command. In a more realistic realm, the limitations of the centrally directed initiatives of the Congress and the Atomic Energy Commission (AEC) and later the Department of Energy and the Nuclear Regulatory Commission, with respect to the deployment of nuclear power systems, were dramatically illustrated by the Three Mile Island accident.

22. The post-World War II qualitative arms race has inspired new technologies for manufacturing nuclear weapons; for defending and delivering weapons; and for detecting and destroying or deterring hostile use. Herman Kahn (*On Thermonuclear War,* Princeton University Press, 1960) was one of the first strategic analysts to point out that, since World War II, new generations of weapons had been devised, deployed, become key factors in international politics, and been superceded, all without their actually having been used.

23. See, for example, *Science and Public Policy,* a report prepared by John R. Steelman for the President's Scientific Research Board. U.S. Government Printing Office, 1947.

24. Examples bordering on the unfeasible were the Sugar Grove radiotelescope, the C5A cargo plane, and the Anti-Ballistic Missile.

25. The concerns of the Department of Defense and the Congress, that military Command, Control, Communications, and Intelligence functions (C^3I) are so complex that they are prone to collapse under stress, has led to hearings and special studies by the Congress, the General Accounting Office, the Defense Science Board, and other groups.

26. See Bernard Brodie, "The Scientific Strategists," and Albert Wohlstetter, "Strategy and the Natural Scientists," in *Scientists and National Policy Making,* edited by Robert Gilpin and Christopher Wright, Columbia University Press, 1964.

27. The Proposal was analyzed by the National Academy of Sciences in a report, *Organization of Science and Technology for Development* (1970) and in 1977 at the Brookings Institution in a report prepared by Lester Gordon for the Secretary of State. See also the report by Frank Press referred to in Note 2.

28. For a dicussion of the role of the President's Office of Science and Technology with respect to energy and other policy issues, see *Science Advice to the President,* edited by William T. Golden and published as a special issue of *Technology in Society: An International Journal,* Vol. 2, Nos. 1 & 2 (1980). As discussed on page 69 by H. Guyford Stever, in the middle 1960s, the Office prepared broad-scale studies of energy. These are now regarded as prescient, but they received less attention at the time.

29. *The Global 2000 Report to the President: Entering the Twenty-First Century,* A Report Prepared by the Council on Environmental Quality and the Department of State, Gerald O. Barney, Study Director, U.S. Government Printing Office, 1980, 3 volumes.

30. A draft *Report on the Issues Identified in the Nonfuel Minerals Policy Review* was issued by an interagency policy coordinating committee for public review and comment in August 1979. The Domestic Policy Review of Nonfuel Minerals was never completed.

31. Section 3, Declaration of Policy, of the *National Materials and Minerals Policy, Research and Development Act of 1980* (P.L. 96–479).

32. In addition to the *National Science and Technology Policy, Organization, and Priorities Act of 1976* (See note 7), Title V of the *Foreign Relations Authorization Act, Fiscal Year 1979* (P.L. 95–426) is entitled, "Science, Technology, and American Diplomacy." Title V includes pertinent Congressional findings, declares policy, and places certain implementing and reporting responsibilities on the President. The Act also assigns coordinating and oversight responsibilities to the Secretary of State aimed at long range planning for the application of science and technology to foreign policy.

33. The industrial sectors that have been most adversely affected by foreign competition include: consumer electronics, automobiles, steel, and automated machine tools.

34. See pages 141–144 of the article in *Science* by Frank Press, op. cit., for a discussion of the extensive Domestic Policy Review on Industrial Innovation.

35. The matter of international technology transfers among industrialized nations is becoming exceedingly complex as military and civilian technologies become increasingly intertwined. It has been the subject of government studies with respect to the maintenance of secrecy, national security and effective collaboration among allies. See, for example, the Congressional Research Service report, "International Technology Transfers of Technology: An Agenda of National Security Issues" (1977) and President Carter's Report on International Transfers of Technology in response to Section 24 of the International Security Assistance Act of 1977. The effect of transfers on economic competition are also addressed in such resports as, *U.S. Technology Policy: A Draft Study* prepared in March 1977 by Betsey Anker-Johnson, Assistant Secretary of Commerce for Science and Technology, and David B. Chang, Deputy Assistant Secretary for Science and Technology.

PART TWO

Technology And Developing Societies

PART TWO

Techniques
and Developing
Countries

5

Development
Without Tears
Can Science Policy Reverse
the Historical Process?

JOHN D. MONTGOMERY

"Science policy" has been much discussed of late, but rarely examined empirically. At least since the eighteenth century, governments have joined private entrepreneurs in deploying technology to improve the production of goods and services. This deployment has probably been on a rising curve of policy involvement as each country has entered into industrial competition with its predecessors and contemporaries. Some of the consequences are known, and while no adequate measurements exist, the speed of change in the recent versions of the Industrial Revolution has lowered somewhat the social costs of technological modernization as it was first experienced. These improvements have resulted from improved technology and a better understanding of the nature of social costs rather than changes in technology policy. In fact, few government interventions in the name of modernization have aimed at social betterments as such; they have usually taken the form of industrial prestige projects or of military weaponry. Nor have the objectives of science policy followed the values and purposes of the S & T (Science and Technology) community. Science policy has meant getting scientists and technologists to engage in activities that are different from those dictated by

Portions of this chapter were developed for the U.S. Pugwash Committee, the American Academy of Arts and Sciences, and National Academy of Sciences symposium on "Social Values and Technology Choice in an International Context," Racine, Wisconsin, June, 1978.

their own professional norms, but it has not meant reassigning them to serve the cause of equity. Rather, science policy has usually meant Big Science, Military Science, and Prestige Science, perceived as political assets. And it has, in fact, produced knowledge and advances that would otherwise have been delayed. But, except as a by-product, governments have not used science directly to benefit the mass of the people in whose name kings, presidents, and national planners have mobilized it. If the people are better off as a result of science policy, it was not because governments willed it in the industrial revolutions of the nineteenth and early twentieth centuries. It has come about because improving people's lives paid off, politically and economically, for the investors in science.

Science policy may be defined as the allocation or denial of resources (economic and human, tangible and intangible, including leadership and public attention or social prestige) for one of three purposes: (1) to speed the pace of research, the diffusion of new findings, or the adoption of innovation; or (2) to change their course; or (3) to permit them to continue the present course in spite of rising costs, diversionary interests, or increasing difficulties. In the nineteenth century, and in subsequent rehearsals of the Industrial Revolution, governments tended to pursue the first purpose. The Morrill Act, creating land-grant universities, reflects the first purpose. Twentieth-century patronage of science on behalf of space and military research are examples of the second purpose, as are current efforts to produce environment-sparing technologies and to develop new energy sources. Programs assigned to serve the third purpose usually result from bureaucratic and incremental pressures, such as the desire to maintain an establishment or to signal a national commitment.

These purposes call for different policies toward science, which are sometimes contradicting and always controversial. Like other policies, science requires the allocation of resources, sometimes diverting them from other purposes; the benefits it confers on some groups imply injury to others; and the intensity of the resulting sense of loss or gain converts science policy making from a calculus of social gain to a problem in political economy. Speeding the pace of research often calls for an increase in the budget of a particular group of scientists, a circumstance usually involving denial of support to other proposals. When the goals are to serve economic development, the assignment of funds to technological institutes often encounters stiff resistance to scientists at the national universities, for example. The second category of policy, changing the course of innovation, encounters even sharper opposition, since it means discontinuing existing programs (and sometimes even restraining them, as Colombian planners found when they imposed restrictions on the advertising of food products deemed insufficiently nutritious). Since the third policy, incrementalism, tends to offend no one except economic rationalists, it is the prevailing mode in science budget setting. But, rising costs eventually present challenges even to this complacent style of policy making.

The role of science policy has been variously to encourage, stimulate, support, and be a patron of R & D, and to attempt to restrain those of its activities that might be profitable for some individuals, but costly to the larger society. The interesting questions of science policy are when to do which, and at what cost. Science policy thus consists of many pushes and shoves and blocks and tackles, not of a single uniform prescription of progress.

Invention

The hope that science policy can reduce the social costs of future industrial revolutions does not necessarily assume a background of government benevolence. Scientific invention has demonstrably improved the lot of man, moving with each generation at a slightly faster rate while producing somewhat fewer social costs. Politicians sometimes persuade themselves that it pays government (both domestically and internationally) to convert to policy what has hitherto been a by-product of technological advance.

That by-product is not trivial. The quality of life of the average citizen has been enormously improved by applications of science and technology. Gross estimates of the contribution of S & T to the GNP in twentieth-century United States range from 40 percent to 60 percent, and even to 90 percent[1] There is no room for doubt that in the absence of science, the world's population would be quantitatively worse off than it is. If such measures seem too materialistic in the context of the qualitative aspirations of some critics of modernization, equally dramatic improvements in the quality of life, including health and access to knowledge, can be shown to separate this from earlier centuries as a result of scientific advances. The conclusion is inescapable that most of the improvements in the life of the average man over the past two centuries — his clothes, his comforts, his skills, the scope of his ambitions, and his empathy toward still a better life — are products, or by-products, of applications of scientific technology. Medieval civic life may have produced for the Medici or the Bourbons an esthetic better than ours, but even they could not enjoy air conditioning on a sultry August afternoon; the modern commoner has as much access to cultural greatness as kings and electors once did, if he is inclined to take advantage of it.

For countries that are not on the frontiers of discovery, true invention is difficult except in fields that have not already been the objects of R & D in the more advanced countries. But, there are such fields, involving applications made possible by scientific advances elsewhere, but not needed by the countries where the original advances occurred. Identifying and supporting these kinds of R & D are, perhaps, the most promising elements of science policy in the developing countries.

In developing countries, scientific choices among such applications are especially significant because resource constraints render duplication and redundancy far more costly than in competitive industrial societies. The external pressures are all the greater because change is an urgent demand, and the time horizons of decision making are fixed in the far-off future because so many choices will be essentially irreversible. Thus, the need for analyzing the social impact and conflicting values of technological change in these contexts far outruns the capacities for doing so.

To consider the alternative futures of technological choices requires a reappraisal of four issues that dominate our assumptions about the past: the extent to which new and borrowed technology has been the ultimate source of development; the extent to which science policy has influenced the speed and direction of discovery; the possibility that policies regarding *diffusion* of science and technology may have had more influence on the quality of life than has discovery itself; and the more recent conclusion that by changing the processes by which innovations diffuse, governments and international agencies can exercise, especially in the short run, a benign influence on the quality of life in the less developed countries. The hopes implied by these considerations rest on a reconstructed history. Our alternative future is one in which industrial revolutions can avoid the disintegration of the family, the abandonment of the countryside, the malnutrition caused in rural areas by the transfer of commercial food to the cities, the exploitation of child labor in gloomy factories, the splintering effect of class warfare, and all the other costs of rapid industrialization as experienced in the West.

A haunting question often posed to the international community of science historians is whether and to what extent the less developed countries, in their march to modernity, may side-step the first Industrial Revolution, especially by reducing the suffering the rural population and the urban workers underwent during the first generations of the transition. The question implies a healthy skepticism about the recent past and a soupçon of wishful thinking about the near future. It challenges conventional opinions about science and technology, and about economic development, which have been interpreted as accepting suffering as the cost of progress. It reverses, as well, a once-favorite assumption that there is a unitary direction of progress, whether inevitable, or accidental, or purposive. It even asks us to compare our times with a golden age preceding modern technology (although there are a few left who are willing to argue that the average man really lived better in the eighteenth century than he does in the twentieth). Most of all, it requires us to consider how man might be able to combine the best of the past two centuries and avoid the worst. The question has spawned a new profession of science and technology critics worrying about what is "appropriate" for the needs of particular developing countries and how it can be discovered, diffused, and adopted. Are there possible applications of science and technology to the problems of the developing countries that do not impose costs to the poor, or losses to the traditional elites, and that do not enlarge the symptoms of dependency?

Values in Science and Technology

Only a decade or so ago, some scientists were still prepared to argue that their work and their achievements could be ethically neutral. It was convenient to believe that discoveries could be *applied* as rationally (or as serendipitously) as they were *made*. This belief led to the hope that the social choices involved in applying technologies could be made professionally, in the same objective, cool, and detached fashion in which scientists are presumed to work. There would be a clear-cut division of labor, in which scientists would discover and develop new knowledge, and other specialists (e.g., politicians, economists, policy scientists) would accept the responsibility for its social impact.

But, such beliefs are increasingly hard to sustain. If ever social values could have been separated from scientific fact — if a Leonardo da Vinci could once have designed tanks and submarines without worrying about their production and deployment — few believe they can be now, in an age of nuclear weapons, or even of atomic power, food preservatives, insecticides, power steering, high-yielding varieties, supersonic transports, or hair bleaches, each of which has social debits as well as benefits. The problems and the choices are becoming urgent in fields where once the value of progress seemed obvious.

One response to this dilemma has been the development of an intellectual discipline known as technological assessment, or the "choice of technique." As this discipline has progressed, the early intuitions about the "technological misfit"[2] and the need for social and technological "congruence"[3] have given way to more systematic analysis. The issue of "factor proportions" has been applied critically and rigorously to industrial technologies in the hope of achieving an appropriate balance between capital-intensive and labor-intensive methods.[4] Attempts to add other dimensions beyond those of capital and labor intensiveness have followed,[5] and the variety of social considerations introduced into the models of technology assessment have made it the equivalent of a social movement.[6] Each step of the journey has increased the range, if not the power, of the analysis. It would, no doubt, be possible today to attach weights to each human value affected by a proposed technological innovation, and to emerge with a calculus of the potential social costs and benefits of any choice, as seen by the groups most likely to be concerned with it.

But, the movement was arrested almost before it had begun. Unanswered questions are threatening to leave technology assessment a mere intellectual pastime. What should be the denominators of value? How do the group interests add up? To such questions technology assessment has no answer.[7] And, its findings are consequently having little independent effect on ultimate decisions about technological innovation.

There is as yet no convincing basis for fixing the priority among human values; even the taxonomies of value differ. The hope of performing systematic value analysis persists, however, as further investigation produces equivalencies among them. The categories of choice are repeated often enough in the major ethical formulations, and offer enough similarities among the leading theories, to yield plausible comparisons among alternative courses of social action. Discovering a means of satisfying the conflicting values affected by science and technology requires thinking about the social choices that have to be made.

Most ethical systems assume that values lie within the range of individual preference. Individuals can create and make use of social instruments to advance their value status or to prevent their deterioration. Individuals can also reject the social instruments themselves; theoretically they can even opt out of society if the costs it imposes upon their own values outweigh the benefits. Because of such possibilities, society (in its liberal political forms) takes heed of individual challenges to its preferred values as soon as they reach sufficiently uncomfortable levels of intensity. Political leaders recognize, if not a social contract, at least the obligations they owe their constituents in return for continued support. There is a link between how governments use knowledge (and other resources) and their own survival.

In the twentieth century, science and technology provide some of the most important issues of public morality; few of the questions dividing mankind are unaffected by discoveries of new products, new weapons, new means of communication, and new methods of production. Emigration, strikes, guerrilla warfare, and apathy have all been, on occasion, responses to misapplied technology. And, they have also been the occasion for new technological solutions (for example, respectively, through the development of marginal lands, improved industrial safety, better weapons systems, and more pervasive mass communications). The trouble is that such technological "fixes" can also permit a few individuals to manipulate the system still further, usually to the disadvantage of large elements of the populations. Technological innovations that create disadvantages to unrepresented groups in society may thus benefit the sponsors of research and still present a serious ethical problem for scientists.

Because of such conflicts, an important role of governments in all Western traditions has been to seek a viable equilibrium among value-serving social institutions that are in competition over the balance of human values in the community. The process by which this equilibrium is sought requires a continuous search for acceptable relationships among the social institutions involved in an innovation. Governments intervene when the imbalance becomes, for any reason, a political problem. But governments, perhaps especially in the developing countries, do not fully represent the groups most likely to be placed at a disadvantage as a result of developmental change. Effective political action is rarely available to the very social groups that have least control over the development and introduction of technology. The greater a nation's dependence on technology for achieving its collective goals, therefore, the more serious is the potential ethical challenge of scientific activity.

The first step in considering the range of effects technology may have on human values is to identify and classify the potential issue areas. There are many existing formulations of human values that might be used as a starting point. The United Nations Declaration of Human Rights enumerates, perhaps, the most generally accepted list, and it can be used very readily as a basis for a preliminary screening of human values, because each element of the formulation represents an arena in which citizens' claims can be advanced and appraised. Moreover, this statement of values, though somewhat eclectic in origin, can be collapsed into eight fundamental categories[8] to be used in considering the potential effects of science and technology. These categories are security, knowledge, material welfare, psychological and corporal well being, self-expression, community and group loyalty, respect, and morality or rectitude.[9] There is no assumption of uniformity of individual preferences *within* each category; the list is also neutral as to the priorities to be assigned *among* the categories.

In analyzing the effects of technology choices on these values, it is necessary first to acknowledge that each individual aspiration within the range of human values can be affected positively or negatively by scientific discovery and application. Both the evidence for this assumption and the issues it poses may be stated simply:

1. Man's desire for *security* (and its converse, his desire to control others) is affected by all discoveries that change the distribution of power among individuals and nations. Which "side" shall science serve — domination or protection — and under what circumstances?
2. His thirst for *knowledge* (and its converse, his desire for secrecy or exclusive possession) is the source both of scientific endeavor and also of efforts to limit access to its fruits. Who should have access to knowledge?
3. His desire for *material goods* (and its converse, the impulse to surpass the possession of others) is well served by technology, but in excess this desire also inflicts ecological injury suffered by all members of society. Which consumer goods should be supplied, and which restricted?
4. His need for *comfort* is a value that is widely sought and rarely denied to

others, *except* when individual and social needs cause him to inflict punishment or injury upon an enemy. Thus, medical, pharmaceutical, and psychological technology contribute to both health and to torture, drug abuse, and involuntary confessions. What role should scientists perform in the uses society makes of their discoveries?

5. His *self-expressive* and aesthetic desires are served through the exercise of skill and self-discipline. Technology can enhance these qualities through the inventions of new instruments, tools, and media, but it can also replace them through obsolescence and other atrophying innovations. What should the scientific community do when both results are observed?

6. The human values of loyalty, affection, and *community* are associated with the desire "to belong," and they are both aided and undermined by technology. The inventions associated with mass communications have both joined and isolated humanity. Should radio and television ennoble as well as vulgarize? Can the disciplines of industrial society produce community as well as alienation? Are scientists responsible for the uses of the instruments they have created to affect air waves?

7. The values of self-esteem, recognition, and *respect* are sometimes acquired at the ironic costs of denigrating others. They have been both served and challenged by discoveries of psychological and genetic differences among individuals, races, and classes. Should scientists seek ambiguity on these issues, at the cost of manipulating knowledge in order to avoid the risks of bigotry and ethnic stridency?

8. The ultimate restraint value, *rectitude,* may be man's desire to be, and to seem, moral and righteous. It may also be the only value in the policy scientist's panoply that is not directly affected by some phase of science and technology activity, though its theological context has at times been severely threatened by discovery. Should scientists compromise on such discoveries until religion has a chance to catch up?

All of these arenas of value are served by social institutions, each of which in turn is likely to have preferences regarding the development and use of technology.

Recognizing the variety of human aspirations within these categories, psychologists and sociologists have sought to discover patterns among them that appear to be most compatible with scientific or technological standards. This search has enabled them to identify the characteristics of "modern man" and to use these characteristics to specify the social "pre-conditions to modernization."[10] In order to achieve such pre-conditions, technology is expected to introduce the disciplines and values of modern life to any society that aspires to them. But, a dilemma appears as soon as these transitional features appear: the introduction of science and technology either immediately benefits certain individuals or groups whose values are most compatible with the behavior required in "modern society," or else it demands changes in the value structure of individuals and groups that desire to take advantage of the new opportunities for self improvement. Is science and technology, therefore, to serve the privileged few whose values and resources are compatible? Or, on the other hand, is it to manipulate divergent values so that all may take advantage of the opportunities it may create?

The response to this dilemma has been ambiguous. If only a few farmers have benefitted from the Green Revolution, scientists have sought new technologies for those left behind. If an innovation produces a slow popular response, the tendency is to raise the level of government activity, in order to achieve the desired output without mobilizing the public's acceptance of change.[11] In short, during the course of "modernization," values do undergo change, but they do so slowly, and the core is resistant to assaults from reason or from new experience. Individuals prefer to live inconsistently, suffering "cognitive dissonance," rather than to make conscious realignments in their value preferences. Science policy must, therefore, either ignore or seek to accommodate the values of individuals (taken collectively) who remain holdouts against change or who are injured by it. When this problem becomes a conscious issue of science policy, the usual outcome is an attempt to create mechanisms for discovering the sources of resistance, and if the conflict appears irreconcilable, to seek new, "trade-off" options. More often, however, the response is: Full speed ahead; Devil take the hindmost.

Appropriate Technology and Beyond

The first intellectual task accepted by scientists in confronting the technological imperative has been to find ways of accommodating the values of modernizing societies by identifying an "appropriate technology" that contains the losses imposed by change. It is a minimalist approach, seeking the technology that inflicts the smallest injury, or confers the greatest benefits, on most members of a society. This utilitarian formula seeks to minimize unemployment, for example, or the outflow of capital, or the damage to the inhabited environment. It can provide sophisticated answers by applying such tests as Paretian optimality to social choice (for example, seeking the least injury in the location of a nuclear power plant).[12] In cases where a choice is irreversible, the effort to reduce community protests to an absolute minimum justifies a substantial analytical effort on the part of scientists and technicians participating in the planning process.

The models now available for such analysis are sophisticated enough to incorporate almost any combination of a community's value preferences into the equation of choice. Unfortunately, however, identifying the least injury to the most people is not sufficient. If it were, defining the minimalist position to social choice would be little more than a problem of good staff work. But there is no staff objective enough, or wise enough, or influential enough, to separate these controversial issues from the decision maker's ethical responsibilities.

The social utility of even good staff work is limited by the attentiveness and commitment of the decision maker to whom it is presented. Visitors to almost any national ministry in the Third World have observed and sympathized with the plight of its top professional officials: They are harassed and over-burdened by unfamiliar problems, which force them to engage in constant negotiations with counterparts and colleagues. Their opportunities for any serious review of long-term problems depend on the political volatility of their environment and competing demands on the slim ranks of talented technicians available to take over where they leave off. The principal cause of the frustrations so frequently encountered by staff analysts whose best work is ignored at decision-making time is not so much that their ideas are rejected, as that they are not even considered.

One result of this official neglect is an underground network of good but unapplied staff papers. An unsuccessful program developed in one country often appears as the basis for a working paper in another country, and its most advanced version may be finally adopted in a third. International consultants and policy scientists serve as the conveyor and preservers of these untested staff papers until their ideas, approaches, and methodologies develop a life of their own. Such recognition is not as good as adoption, but it is better than oblivion. Each rehearsal of an untried proposal gains something by way of renewed assurance and a professional constituency; decision makers have been known to adopt, on the rebound from other countries, proposals that originated from their own staffs and then reached them again via the international circuit. In the post-Xerox age, policy analysts never know where their work will wind up. Thus participation by scientists in studies of the social consequences of their technologies is a casting of bread upon the waters.

The reason why this underground of unsuccessful staff papers is so important in social choice is that it significantly augments the capacity of decision makers in the Third World to consider new options. Decision making in most developing countries suffers from the small and narrow bases of the national elite structure, which lacks both versatility and diversity. Social and political organizations in the western democracies provide policy guidance, sanctions and restrictions, and replacements and support to decision-making elites, but in the developing countries they do not function effectively as instruments of social choice because their decisions are rarely made in a public forum. As a result, the staff work in these countries, with their widely different perspectives and traditions, provides for the decision-making process something of the same service of proposal generating and coalition building that different interest groups do in the Western democracies. There are not circulating elites in the Third World so much as there is an elite circuit for new ideas and programs. It would be instructive to trace the flow of innovation ideas about rural development from East Pakistan to Tanzania, and from Indonesia to the Philippines, or even from Maoist China to Ethiopia, by studying the proposals that were unsuccessful in the short run. They would tell a great deal about how social choices are made.

Elite decision making and the international flow of staff work affect the role of the scientists in the formulation of technology choices. Most countries have, it is true, social institutions that embody the values enumerated above, all of which are potential patrons or opponents of applied science and technology. There are, respectively:

1. armies and police forces to provide for collective and individual security;
2. universities to protect higher learning and the search for knowledge;
3. banks, insurance companies, and corporations to serve as custodians and mobilizers of wealth to produce consumer goods;
4. hospitals and medical professionals to restore health and well-being to the community;
5. museums, folk traditions, and trade schools to encourage and promote self-expression;
6. unions, farmers' associations, extended families, and ethnic organizations to serve the needs of workers and other citizens who desire communal association;
7. courts to protect individual rights and uphold the privileges conferred by law; and
8. churches to symbolize righteousness and remind members of their moral obligations.

The full panoply of value-serving institutions is on display in every country where science itself is allowed to exist. But these institutions, often small in proportion to the population, act in isolation from each other, each pursuing its own ends and serving its own vision of social values, rarely interacting with other institutions in search of higher resolutions of values in conflict, and even less frequently becoming involved in the technological issues of social choice that have dominated American and European discussions of science policy in recent years. Scientists concerned with social choices find little intellectual guidance or organizational support from the value-serving institutions of the Third World.

Most countries, even among the least developed, have established political processes for considering social values and assessing national priorities. The concept of the public interest has not emerged as a significant force affecting these processes, however; the succession of private interests that attempt to colonize the public sector in the developing countries has not provided an arena in which values clash so much as a battleground for different claimants to power.[13] It is rare that issues of science policy are identified at all in the political process, even when major social costs and benefits are at stake. Instead, these issues are characteristically decided in response to pressures from the institutions and are organized around value sectors. For these reasons, national formulations of social values in the developing countries tend to be conventional, limited in scope, and lacking in precision, reflecting elite judgments rather than abiding social or national purposes.

Because, in addition, neither the institutional representation of values, nor the political processes of decision making permit scientists to inject their views of social choice into the design and application of technology, the task of scientists in developing countries who must consider problems of social choice is on the whole more difficult than that of their counterparts in the Western industrialized countries. They find political decision makers inaccessible through the use of "normal" staff work; they can locate few active organized interest groups to which they can express their preferences and reservations; the social institutions that articulate values in the Western democracies are isolated and narrowly focused in the developing countries; there is no tradition of public interest to which they can relate their concerns; and official formulations of national goals and purposes are transitory and often inapplicable to the issues and dilemmas posed by scientific advance. If their perceptions of the costs and benefits of a new technology are to come to the attention of public and private officials, scientists will have to adopt roles and take upon themselves responsibilities that are heavier and more active than those of their counterparts in Europe and America.

Scientists as Activists

Although most developing countries are elite-dominated to a much greater degree than are Western industrialized societies, government decisions are not entirely determined by elite self-interests. It is true that coup leaders tend to leave office rich, if they do so at all; that many government-distributed benefits go directly or indirectly to their allies and to counter-elites who have to be placated; and that the rural and urban poor in most of Asia, Africa, and Latin America are under-represented in the councils of government and cannot mobilize the efforts of scientists and technicians who are discovering or applying new knowledge in their community. But, in spite of these disturbing patterns of policy making in the developing countries (and, indeed, to a lesser extent in the industrialized countries as well), there are also trends in the direction of responsible social choices that are increasingly evident. There are countervailing forces to redirect discovery and investment toward social goals that transcend the immediate self-interests of the present elites.

Part of the reason for the rise in responsible social choice is that political elites do not always recognize the crucial decisions that determine the future directions of their society; they prefer to leave long-term problems to scientists and technicians. Important social choices are, therefore, made by international and domestic "experts" concerned with issues that are not perceived by political leaders as central to their interests. Scientists are increasingly providing the sources of long-term and equity considerations in national policies and economic development plans. The international influences at work on these scientists range from statements of human rights and exhortations in the United Nations' charter and other documents, to specific interventions by way of research and agriculture to improve the quality and supply of food. Scientists themselves have links to the international community that draw their attention to their colleagues' increasing concerns over population, nutrition, and ecology.

Thus, praetorian regimes in which scientists are allowed to make politically unimportant decisions, provide some of the best examples of national planning involving major social purposes in the generation and application of knowledge. They may lead nowhere in their immediate political context (Who can tell?) but still provide intellectual stimulation to other countries. Even Nicaragua, recently torn by civil strife, had formulated a food and nutrition plan in 1977, for example, that linked government programs to scientific research in a comprehensive, integrated program responding to the needs of the poor. The plan identified three elements of the population considered the most vulnerable to the effects of malnutrition as follows: (1) children under five who are members of low-income families; (2) pregnant and lactating women; (3) low-income families, especially in rural areas, at risk because of specific nutritional deficiencies like iron, iodine, and Vitamin A. The programs to be undertaken in reaching these target groups were to involve workers in both the public and private sectors of health, agriculture, and education. They called for resources from international sources, including the United States, and from regional nutrition and agricultural centers, to be mobilized at the community and national levels. The technologies required the use of Vitamin A fortification, salt iodization, iron enrichment, and media campaigns, as well as the development and production of appropriate blended food products, the design of integrated rural development programs, and the analysis of the food-distribution system. The plan provided for nutrition surveys and recurrent surveillance activities and the evaluation of program impacts on the intended beneficiaries. It was, in short, an integrated assault on the problem, involving the application of all relevant current knowledge, the development of new technologies, the conduct of social and food research, and the coordination of a massive administrative operation.

All of these aspirations were generated at the technical level in Nicaragua, strongly influenced by international planners and the expectation of a substantial development loan. These technical and international actors were crucial; the plan had to be developed and promoted in the face of political indifference to the social values involved. The strategy of the scientists and planners was simple: to demonstrate the coherence and technical and financial feasibility of a balanced approach, and in the process to develop enough bureaucratic and international support to gain the attention and sympathy of political decision makers. This political support, of course, did not come easily. Earlier efforts at nutrition programming had been fragmented and unimpressive: A 1969 law requiring the iodization of salt had never been implemented; a 1972 bill to establish a national nutrition council was never passed; a 1973 announcement that U.S. subsidized food programs (PL 480, Title 2) were to be phased out for want of commitment to nutrition was ignored; even the 1975–79 national plan had made only scanty reference to the problems of malnutrition. But mounting interest in the issues of income redistribution and rural development had begun to dominate the discussions in the national ministries, and a sense of jurisdictional competence, together with the possibilities of reinforced, cooperative efforts, encouraged the scientists and technicians to begin the process of constituency building, to seek international loans, and to develop commitments from the chief lawmakers to a coherent program.[14] Could a dictator use such programs as a claim to legitimacy? That was the gamble the technicians and planners were taking.

The process of coalition building that was followed in Nicaragua was typical of elite-based efforts to build national programs of social choice around new clusters of emergent technologies. It was not a political ploy to preserve a faltering regime. An official paper described the effort as follows:

> The Policy is not expected to create support for a National Food and Nutrition Program where none existed before, but rather, to crystallize the diverse public and private support that exists, broaden its base, and facilitate its institutionalization. . . .

Links with private physicians, educators, administrators, local officials, and social leaders would enhance the professional contributions of the scientists and technicians who had developed the plan.[15]

This kind of direct action seems to be an increasingly important preoccupation of scientists and technicians. Even those doing research within narrowly defined fields usually have to work behind the scenes to influence social choices arising from their findings. They have to make choices: For whatever the objective of a scientific endeavor, its potential use involves conflicts in human values. Thus, the roles of the scientist as discoverer or as technician are seldom isolated; with what other actor or actors should he seek alliance? Each value arena offers the prospect of coalition partners in policy making whose interests may be adverse or helpful to larger segments of the society. For example, the armorer who serves the military client can threaten the society he is supposed to protect; even the "pure" scientist can threaten the establishment with his findings, or help traditional religious leaders interpret new technologies in accordance with old folkways. The industrial technician working in a government research institute can concentrate on projects beneficial to large corporate interests or on small-scale projects or on consumer welfare; and such dilemmas, involving selection among clients and partners, permeate the process of social change.

The Diffusion of Influence

Few scientific endeavors can escape consequences that some humane scientists participating in them would prefer to avoid. These consequences are not the goal of scientific endeavor. They occur because organized knowledge changes both the ways groups of people do things and the ways they perceive them. The changes are not uniform, and they are not fully predictable. Each sponsor of scientific research — whether the army, the corporation, the research and development laboratory, or the university — seeks to gain from the mastery of new knowledge. They are not by virtue of their sponsorship the only coalition partners scientists seek in issues of social choice. Sponsors may perceive advantages in withholding knowledge, as well as in using it (even universities may try to keep armies or police investigators or commercial exploiters or competing researchers from gaining access to some knowledge). Thus, the scientist cannot escape a role in social choice by efficiently serving a sponsor, even though its role in social change is supposedly benign. By gaining access to knowledge, the scientist assumes responsibilities that go beyond those of the uninformed citizen. Loyal alignment to institutions serving good social ends does not free him from that responsibility.

A convenient definition of the role of scientists in development is the distinction among the functions of "generation," "diffusion," and "application" of knowledge,[16] which must be linked to form a triangle of service to social ends, since each function is necessary but is insufficient in itself as an instrument of man's betterment. But, each function can also be diverted away from social purpose; each can be misused if aligned too faithfully with national ends defined by a political leadership; and each can diverge from the common welfare if its orientation to international purposes result in the neglect of domestic priorities. Scientific policy involves the search for an appropriate balance of resources devoted to the generation, application, and diffusion of knowledge in a given setting. Both national and international science communities are aware that effective science policy requires a balancing among the links to foreign and internal suppliers and users.[17]

This definition of functions, however, does not afford much guidance to individual scientists confronted with the different problems of social choice. And scientists need guidance: expertise in atomic physics, ovulation cycles, or weather modification is not a guarantee of social wisdom. The analytics suggested here is a first step in structuring the inquiry into social choices. But even mastering it is not sufficient: a more active role is often necessary. For the analysis of an appropriately active role, a different set of definitions is necessary. Scientists whose knowledge confers special responsibilities beyond those of the citizen affected by a forthcoming social choice can discharge those responsibilities in one of six ways. They can serve as *advocates* (entering into the policy arena directly); *advisers* (analyzing the consequences of different courses of actions for different groups); *decision makers* (in which case they themselves, not their sponsors, accept the responsibility for the uses of their science); *critics* (the advocates of alternate choices); *conspirators* (organizers of opposition forces); or *activists* (leaders of demonstrations and other protest movements against a given use of scientific knowledge).

Dividing one's activity among these roles is the element in social choice that goes beyond minimalist staff work. The calculus of social consequences — whether through technology assessment, utilitarianism, Paretian optimality, or social justice theory — is only the first stage in the exercise of social responsibility. The information derived from this form of staff work is insufficient without the exercise of one of the more active roles. The scientist may not have to play such a role very often, but where his own work is involved, he cannot evade it unless he is willing to allow the product of his hands to serve the immediate self-interest of the strongest and most powerful of his contemporaries.

Development Without Tears

Reducing the harmful effects of a malignant science policy that ignores social values in the interest of an elite does not provide protection against injuries caused by autonomous scientific effort, however independent it may be of the preferences of unscrupulous or short-sighted politicians and planners. According to laissez-faire theorists, technological discovery is fueled by its own vital force, responding to the effects of private emulation, or greed, and not to the preferences of economic planners, political leaders, or foreign aid donors. The appropriate role of government, it would follow, is to control expressions of greed that injure the public, and to restrain forms of emulation that offend higher social values, letting technological innovation follow its own course.

Closer examination is damaging to this interpretation. When the effects of innovations in science and technology to development are disaggregated, it appears that they are (1) different in different countries, even those enjoying approximately the same standard of living; (2) different in different sectors within countries, permitting modern industry to coexist, for example, with stagnant agriculture; and (3) different at different historical periods, so that the Industrial Revolution looks quite different as it appeared in Britain, Prussia, the United States, the Soviet Union, Japan, Korea, and Singapore. The implication is that science policy, or something like it, has intervened to cause the difference, but in no necessarily discernible pattern.

Moreover, it is probably still true that greatest returns on investments in science come from the diffusion, not the generation, of technology. The trouble is that diffusion is not what scientific research institutions do best (or in many cases at all). My own studies of how science policy agencies in Asia and Latin America perceive their functions confirmed this expectation. There are 174 science policy bodies listed in the UNESCO directories for Asia and Latin America. In their descriptions of their own activities, 84 percent perceived their roles as generating and devising technological systems; and 65 percent expressed interest in institutions that are involved in introducing S & T into the society; but only 28 percent perceived it appropriate to concern themselves officially with the needs and responses of user or client groups whom they were presumably expected to serve. At regional and country levels, there are variations in these perceptions, but the ranking and proportion are not significantly different.

One reason for these limited role perceptions is the narrow range of links connecting science policy bodies and user groups. These linkages are of six kinds: public-private (for planning, funding, and policy making); interministerial (for intersectoral cooperation); vertical, to clients (for implementation, marketing, and/or adoption); interministerial (to encourage and discourage different kinds of competition and collaboration); regional (for inter-country research and dissemination, especially in agriculture and medicine); and international (import/export and exchange functions).[18] In examining these linkages, it is clear that international transactions are more important (i.e., more frequent) than any domestic linkage; that intersectoral linkages are rather weak (fewer than half of the countries listed have established such relationships as an official relationship of science policy bodies); and that the vertical, direct linkages to clients are rarely encountered among government science bodies.

Diffusion of science and technology to farm and urban users is not a function that governmental science bodies are disposed, or even equalified, to perform directly. The UN *World Plan of Action in Science and Technology* described the appropriate functions for these agencies, but did not list among them any developmental outreach efforts or responsibilities for technology diffusion.[19] If science/technology diffusion is so important for development, other institutions and intermediaries than science policy bodies and public research laboratories will have to be created.

Technology development that takes place through a series of "accidents" may result in successful diffusion if the innovation happens to coincide with an existing demand. But such results are rare and serendipitous. Other consequences are more frequent:

1. the new development is aborted early; or
2. the innovator has to engage in a marketing push to recover his investments; or, most seriously of all,
3. the users make inappropriate choices for want of known alternatives.

These mishaps are commonplace in the commercial market, and their effects on both producers and consumers are often cited as flaws in the economic system. To the extent that competition eventually offers correctives to distorted demand-supply relationships, not much is lost by these innovative failures. But these correctives are usually not available for "public" goods like agricultural technology, health delivery systems, or educational services. In the developing countries, especially in situations where government decisions replace market choices even regarding "private" goods, the luxury of "inappropriate" technology is a serious drain on available resources.

Current criticism of "inappropriate technologies" may seem faddish, but it has supplied us with a generally accepted inventory of these costs:

— unemployment
— underemployment
— people deprived of health care and schooling because the delivery systems are unable to serve them
— irrigated farms producing one crop a year instead of two or three because water is distributed wastefully or inequitably
— enclave high-technology industries producing subsidized goods for export when the domestic market is undersupplied with essentials.

Even moderate critics conveniently lump all such distortions in the use of S & T for human welfare together as the products of over-eager technology transfer, "inappropriate" technology that should be replaced with "appropriate" or "intermediate" technologies because "small is beautiful" and "simple is optimal."[20] Radical critics regard the evils as the inevitable consequence of a market-based economy. Advocates of science policy believe that judicious use of government resources can adjust the decisions that lead to these distortions.

Science policy as a solution has both international and domestic dimensions. The belief that better technologies can be invented anywhere, and diffused to replace current ones, has led to the establishment of research institutions in Britain and the United States, funded out of international assistance budgets. Regional agricultural research institutes have sprung up in Asia, Central America, and Africa to study tropical conditions and crops and to develop approaches to small farm mechanization. Their task is to create alternatives to inappropriateness. A growth industry in economics has developed out of analyzing employment-intensiveness or energy-utilization of alternative technologies, hoping to stave off in developing countries the impulse to inappropriateness. New products and processes have already emerged from these efforts, along with new insights into the technologic decisions open to development planners and administrators.

But this approach ignores the still greater potential capabilities of the S & T communities in Asia, Africa, Latin America, and the Middle East. The science bodies in those regions are only indirectly influenced by improvements in "donor" perceptions and practices. To them, the "intermediate technology" movement is only a ripple in that sea, a fad of the rich that may actually aggravate technological dependency.

Science in developing countries is clearly dependent on American and European research, in spite of the dramatic rise in numbers of qualified scientists and technicians and the rapid emplacement of university and other research facilities. This dependency is reflected in the professional aspirations of scientists (to visit Western colleagues and laboratories and present papers and articles in Western journals), as much as in the unidirectional flow of technology (as indicated by patent fees and licensing arrangements). Local scientists and technologists are linked more strongly to the international community than to their own societies.

The link between client demands and technology generation and diffusion, one of the most powerful sources of support for scientific and technological endeavor in the United States, is so deficient in the less developed countries that much research that goes on there is still unrelated to the principal concerns of the population, in spite of exhortations from the international community.

Improving these linkages is a serious requirement of current science policy. Governments have themselves been serving as patrons of whatever R & D is going on, but in recent years they have begun experimenting with the creation of private intermediaries between technological institutions and user groups. Governments as clients of scientific research are too far removed from the users of technology to influence the choice of problems intelligently on their behalf or to move the course of discovery toward their needs. New technologies succeeded in creating industries in Taiwan and Korea only after the governments tapered off their direction and control of research and turned marketing and dissemination over to private and semi-public institutions. In Latin America, innovations have been "raffled off" to avoid their being exploited exclusively by government laboratories. But private hands are not totally trusted either: Andean Market countries now closely monitor licenses for international transfers of technology, with view to providing market linkages to domestic clients whom the planners (not the foreign corporations) wish to serve. The optimal sponsors of research and the diffusion of technology seem, in most cases, to be an informal, or even competitive, partnership between government and private entrepreneurs.

Recent efforts to avoid direct government sponsorship of research have provided some ingenious organizational inventions:

1. experimental partnerships between public agencies and private industry in the sponsorship of research on problems of interest to both partners. Examples: shipbuilding in Korea; petro-chemistry in Brazil;
2. off-loading of development and diffusion investments to private or semi-public agencies after basic research had established the feasibility of a new product. Examples: agro-industry and pisciculture in Taiwan;
3. regulation and selective licensing procedures to restrain technological innovations of doubtful social benefit and encourage local investment in others as import substituting ventures. Examples: controls on advertising of infant foods that might discourage breast-feeding, and applications of protein texturization and other techniques to domestic food products, in several Latin American countries;
4. incentives to private and cooperative investment in R & D, including public bonds, tax advantages, mixed and semi-public enterprise arrangements, and the sale of stocks in public corporations, with the intention of creating strong direct links between S & T activity and the market.

All such efforts may be regarded as means of replacing supply-dominated styles of technological transfer and diffusion.

Regulation and restraint of foreign competition are not sufficient to establish the needed linkages between local innovators and local users. If governments merely replace international technology donors, without creating institutions that will relate their own protected efforts to local needs and development objectives, the resulting technological monopolies will probably be no more benevolent, though certainly less efficient, than those of current multinational corporations.

The primary lesson of recent successes and failures in the policies addressing innovation-to-user is the need for studying in detail the linkages of incentives, organizations, and transactions that run from the clinic, the laboratory, and the workbench, to the homes and fields where most of the world's population lives and works. Concentrating on such linkages will not eliminate the costs of the latter-day industrial revolutions, but it will facilitate awareness of those costs and stimulate efforts to reduce or offset them.

Notes

1. These varied estimates were obtained from the following sources: R. Solow, "Technical Change and the Aggregate Production Function," *Review of Economic Statistics,* vol. 39, (August, 1957), p. 312; and E.F. Denison, *Sources of Economic Growth in the U.S.,* (New York: Committee for Economic Development, 1962).

2. Walsh McDermott, "Modern Medicine and the Demographic Disease, Pattern of Overly Traditional Societies: A Technological Misfit," *Education,* Vol. 41, No. 9 (Sept. 1966).

3. John D. Montgomery, "The Challenge of Change," *International Development Review,* Vol. IX, No. 1, (March 1967).

4. Richard S. Eckaus, "The Factor Proportions Problem in Underdeveloped Areas," *American Economic Review,* XIV:4 (Sept. 1955), L:2, May 1960); A. S. Bhalla, *Technology and Assessment in Industry* (Geneva: ILO, 1975).

5. John D. Montgomery, *Technology and Civic Life: Making and Implementing Development Decisions* (Cambridge, Mass.: MIT Press, 1974), ch. 5 and 6.

6. Francois Hetman, *Society and the Assessment of Technology* (Paris: 1973).

7. A. A. Fouad, "Choice of Technological Models by Developing Countries," and Gustav Ranis, "Appropriate Technology for the Developing Countries: A View from the Industrialized Nations," papers presented to symposium on Social Values and Technology Choice in an International Context, U.S. Pugwash Committee, American Academy of Arts and Sciences and NAS, June 1978.

8. Myres S. McDougal, Harold D. Lasswell, and Lung-chu Chen, "Nationality and Human Rights: The Protection of the Individual in External Arenas," 83 *Yale L.J.* 900–998 (1974); Harold D. Lasswell, *Preview of the Policy Sciences* (N.Y.: Elsevier, 1972).

9. These eight values correspond to a similar list used in the policy sciences: Power, Enlightenment, Wealth, Well-being, Skill, Affection, Respect, and Rectitude.

10. Alex Inkeles, *Becoming Modern: Individual Change in Six Developing Countries,* Alex Inkeles and David H. Smith (Cambridge, Mass: Harvard Univ. Press, 1974); Daniel Lerner, *The Passing of a Traditional Society: Modernizing the Middle East,* Sept. 1969 (New York: Free Press of Glencoe, 1964).

11. Montgomery, *Technology and Civic Life,* cited, ch. 3.

12. A few years ago, a group of Harvard scientists identified a geographical site for a nuclear power plant to which no identifiable elements of the community objected. Jacques Cros, *Power Plant Siting: A Paretian Environmental Approach,* Discussion Paper 74–4, Harvard University Environmental Systems Program, August 1974.

13. See the author's "Public Interest in the Ideology of National Development," in Carl J. Friedrich, ed., *The Public Interest,* Nomos V (New York: Atherton, 1962).

14. The loan was held up in Congress because of Nicaragua's human rights record, then finally approved because of its exceptional merits as an application of technology to a major poverty problem.

15. "Population Policies as Social Experiments," in John D. Montgomery, Joel Migdal, and Harold Lasswell, *Patterns of Policy: Comparative and Longitudinal Studies of Population Events* (Rutgers, N.J.: Transaction Press, 1979).

16. Jorge A. Sabado, "Science and Technology in the Future Development of Latin America," report presented to World Order Models Conference, Bellagio, Italy, 1968.

17. John D. Montgomery, "Science Policy and Development Programs: Organizing Science for Government Action, in *World Development,* Vol. 2, Nos. 4 and 5, April–May 1974.

18. Ibid., p. 65.

19. *U.N. World Plan of Action* (New York: 1971).

20. See, for example, Denis Goulet, *The Uncertain Promise: Value Conflicts in Technology Transfer,* (New York, IDOC/North America, in cooperation with the Overseas

Development Council, Washington, D.C., 1977); E.E. Schumacher, *Small is Beautiful: Economics as if People Mattered,* (New York: Harper & Row, 1973); and Robert Chambers, *Planning for Rural Areas in East Africa: Experience and Prescriptions,* (Nairobi University, Institute for Development Studies, Discussion Paper #119, 1971).

6

Science and Technology Policy: A View from the Periphery

FABIO S. ERBER

The Importance of the Policy: the Change in Evaluation

The importance of science and technology in our present-day world is well-known — it has been sung in prose and verse and decried with the same intensity by all breeds of analysts.

However, although the influence of science and technology is felt world-wide, their production is heavily concentrated: whether one uses measures of inputs (R & D expenditures, qualified scientists and engineers) or outputs (patents, papers), one finds that the so-called "less developed countries" play a minor role in the international production of science and technology, lower than, for instance, their participation in international trade and international industrial production, as can be seen in Table 6.1.[1]

The recognition that the participation of the LDCS in the international system of science and technology almost exclusively as users of science and technology produced in the advanced countries may pose serious problems to the development of the former group of countries is rather recent.

In all likelihood anyone taking a course in Development Economics in the late 1950s or early 1960s would be told that one of the advantages of the "less developed" countries was the possibility they had of relying upon the

TABLE 6.1   World Distribution of R & D Expenditures, Industrial Value
Added and Manufacturers' Trade — in % — 1973

Countries	R & D	Ind. V. A.	Trade
Less Developed	2.9	7.6	19.0
Developed	97.1		81.0
Socialist Econ.	30.6	24.3	10.0
Market Econ.	66.5	68.1	71.0
World Total	100.0	100.0	100.0

Sources:
1. R & D Expenditures: J. Annerstedt — "World R & D Survey."
2. Industrial Value Added: P. Vuskovic — "America Latinate ante Nuevos Terminos de la Division Internacional del Trabajo," Economia de America Latina, March, 1979.
3. International Trade: Nacional Financiera S.A.: "Mexico: una Estrategia para Desarollar la Industria de Bienes de Capital," NAFINSA/UNIDO, 1977.

stock of scientific and technological knowledge of the "developed" countries.

Such reliance, it was said, would allow great increases in productivity in the developing countries, especially in agriculture, making up one of the main elements of the "economics of the take-off." Industrial technology would be brought to the economic system by foreign investment and by "modern" local entrepreneurs and, gradually and "naturally," a local scientific and technological capability would emerge in the developing countries, following a pattern similar to the "developed" countries. The main obstacles in this path of development lay in the "traditional" values of the developing countries, especially of their elites, and in the lack of "qualified" personnel. Therefore, the priorities for government action were to "sensitize" the local elites (including the budding industrial entrepreneurs) to the value of "modern" technology, to educate peasants and workers so that they could use the important technology and, last but not least, to facilitate the entry of foreign investment. The development of a local scientific and technical capacity going beyond the capacity to operate foreign technologies should rank low in government policies.

Sometimes attention was paid to some difficulties of importing foreign technology, especially as regards scale problems — the markets of developing countries being much smaller than the developed countries — and the consequences of using labor-saving techniques in countries with large populations as the developing countries, often already facing problems of employment. Nonetheless, it was often thought that resorting to "older vintages" of the advanced countries' technologies, coupled to a proper pricing of "capital" and "labor" would provide an adequate answer to such problems.

Although such views are still widely held in both the developed countries and the developing countries (thanks in part to the conservatism of the education system), since the late 1960s there has been a kind of backlash: a considerable amount of research has thrown light on the disadvantages for the developing countries of relying upon science, and especially the technology of the advanced countries. At the same time, several developing countries, especially the more industrialized ones such as Argentina, Brazil, India and Mexico, increased their commitment to develop a local scientific and technological capability, often with an explicit objective of achieving their relative autonomy in such fields.

From such research and such policy-making experience a kind of dialectical synthesis has emerged: While still recognizing the need of policies of support of science and technology in the peripheral countries, we have come a long way from the "naif optimism" (Cooper, 1973) of the late 1960s and early 1970s, when the accent of the policies fell almost exclusively on the provision of qualified scientific and technological manpower, as well as from the subsequent "systems approach," which gave as a solution the establishment of "proper" institutional links between the "science and technology system" and the "productive system." It is probably fair to say that presently policy makers and researchers have gained, as a result of their experience, a better understanding not only of the rationale of science and technology policy in the periphery, but also of the great difficulties the economic and political structures of such societies — and of the international system in which they are inserted — impose on such policies.

The rest of this paper delves into the two points mentioned above: the rationale of a policy of greater technological autonomy, and the limits of such policy in the developing countries. But, before entering this "selva oscura" a few caveats are in order: first, many of the problems of science policy per se, such as the development of a scientific community in its stricter sociological sense, are touched upon tangentially only — the focus here is mainly on technology policy. Second, the analysis deals only with capitalist economies, partly because the author is better acquainted with the Latin American literature, and partly because tackling the issues of a socialist rationale would demand more time and space than are available here.

The Rationale of State Action for Technological Development

Reproduction and Enlargement of the Economic Conditions for Accumulation

One of the basic functions of the state in a capitalist economy is to guarantee the conditions of continuity of the process of accumulation. This function has led most states to go beyond their classical regulatory activities and to act as a direct supplier of goods and services, especially in sectors where such supply is essential for the process of capital accumulation but where the conditions are such that they are not attractive to private enterprises.

The activities of R & D[2] seem to be a case in point: Despite their importance for the process of accumulation, the conditions of production and appropriation of knowledge are such that private enterprises tend to invest in such activities less than what would be desirable from the point of view of the economy as a whole. This condition would justify the intervention of the state in the production of such knowledge, either directly or by granting additional incentives to the firms performing R & D activities.

Part of such sub-investment is due to the specific characteristics of the process of production of scientific and technological knowledge, and of the conditions ruling their appropriation, characteristics common to any capitalist economy. But another part — especially relevant for the peripheral economies — is due to the conditions of insertion of the national economies in the international system.

However, scientific and technological knowledge is not a homogeneous "factor of production" — it is the result of a complex and differentiated set of activities and its importance varies for different economic sectors. Therefore, the importance of state support to science and technology for the general process of capital accumulation will vary according to the sectors which lead the process of economic growth and which probably are the politically hegemonic fraction within the state.

In fact, since the several classes and groups that compose society are differentially represented within the state in accordance with their economic and political strength, the state action will reflect such differences, leading to different science and technology policies in accord with the economic and political structure of the society.

For analytical purposes we shall first treat the national economy as if it were a "closed economy."[3] In the following part, we shall treat it within the context of the international system.

State Intervention in a Closed Economy

Enterprises investing in R & D face three types of uncertainty: a technical uncertainty, in the strict sense, that they will be able to develop the products and/or processes with the desired characteristics; a techno-economic uncertainty that they will be able to produce the products and processes in conditions of quality, price, and time competitive *vis-à-vis* their competitors and purchasers; and a financial uncertainty, especially of "gambler's ruin"[4] when the firm concentrates investment in a few projects.

The uncertainties mentioned above will vary, of course, according to the conditions of the firm *vis-à-vis* the product or process to be developed, its competitors, and consumers. The uncertainty tends to be higher for products and processes that imply a significant departure from the previous experience of the firm, and it will be higher still if there are no external sources of information that the enterprise can use, such as information systems, research institutes, universities, etc. That is, the uncertainty increases when the firm is innovating in social terms.

In the case of the peripheral countries, these points hold a special significance. The importance of experience and learning-by-doing in scientific and technological work suggest to entrepreneurs and policy makers alike the desirability of some caution in the more voluntaristic proposals of great leaps forward in technological development, especially since the research on innovation indicates that such experience is *social* experience in two senses.

First, the scientific and technological activities of enterprises are based upon the activities of a network of other institutions, such as universities, research institutes, information systems, etc. Although "innovating entrepreneurs," full of "animal spirits," may be a necessary condition for technological development, they are not a sufficient condition; they require the support of an extended division of labor, itself the result of a long period of capital accumulation. Where, as in the developing countries, such scientific and technical infrastructure is poorly developed, the innovating enterprise must internalize several activities that in other economies are performed by other institutions, thus increasing its risk. Second the uncertainties mentioned above apply to any new sector and not only to the highly technology-intensive ones; the degree of novelty is dependent upon previously accumulated social experience. It is worth stressing that such experience sometimes disappears over time — enterprises close down, designers die, etc. — a point often overlooked in the discussion of "intermediate technologies" by those who suggest the developing countries should use the old technologies of the advanced countries. Such use would often require considerable technical capacity.

The uncertainty inherent in R & D investments which is especially high for basic research activities, is amplified by "imperfections" in two types of markets — insurance and capital. The lack of insurance markets, which would "price" the risks of innovation,[5] and the lack of risk capital for this type of investment are especially pronounced in the peripheral countries,[6] except where the state supplies such funds.

The lack of risk capital is especially important for industries where the minimum scale of R & D expenditures is high. When such an R & D threshold is followed by high minimum scales of production and marketing, the barriers to entry and the barriers to innovation in such sectors are very high, especially if local markets are small, as is the case for the peripheral countries. Policies of support of R & D in technology-intensive industries, such as electronics, aeronautics, and nuclear energy — justified in part by the lack of risk capital — have often been followed by policies of restructuring the industry by mergers of enterprises, especially in Western Europe (Pavitt et al. 1974).

However, in conditions of imperfect knowledge, risks may be overestimated because of lack of information, including information about the investment of other enterprises, and this imperfect knowledge will compound the deficiencies of the capital and insurance markets and make the enterprises' risk discount greater than the social discount.

In this sense, the action of the state as coordinator of economic activities and provider of information to the individual enterprise about the "state of the arts" in the industry is of critical importance.

Moreover, in social terms, failures in one producer may be compensated by successes in others. Such an approach is, of course, exclusive to the state for each individual producer the failure to develop a commercially and technically appropriate design may be highly damaging, although the state can, as it does in the advanced countries, reduce the consequences of such failures, e.g., by grants (*ibidem*). Furthermore, this "portfolio approach" will tend to be limited to some industries where firms produce a wide variety of goods.

Therefore, the eventual difference between social and privatre risks will depend on the type of product: In capital goods, for instance, it will probably be greater for standard products than for custom-built equipment. It will also depend on the degree of concentration of supply. the fewer the suppliers are in relation to the demand, the greater will be the effects of their failures on other producers or customers.

Finally, since the state acts as a mediator between the interests of the different classes and especially between the demands of the fractions composing the power bloc (Poulantzas, 1968, 1974), its time-discount rate may be different from that of private enterprises (at least some of them) and so will be the evaluation of the time involved in the development of technology. When the state's discount rate is lower, the case for its intervention in favor of local R & D is reinforced.

The combination of costs × uncertainty × time, discussed above, must be counterposed to the difficulties that enterprises face in appropriating the results of scientific and technological activities.

Such difficulties have roots in the scientific and technical labor-process. To a considerable extent, scientific and technological knowledge is person-embodied and is developed by learning-by-doing. When people move from one enterprise to another, they bring along knowledge acquired in the firm where they worked, diffusing such knowledge to other enterprises.

Johnson (1970a) has argued that the existence of externalities in the case of movement of personnel would depend on the organization of the labor market — that labor could pay for its training through lower wages and then be compensated by the firm to which it has moved by higher wages.

However, only under very restrictive assumptions will there be no externalities involved in such movement: not only would perfect labor markets be required, but it would also be required that the costs of training and the increases in productivity of scientists and technologists, because of such training, be perfectly identifiable. As the training in science and technology is based on learning-by-doing, and the production of innovations is highly dependent on personal creativeness, such assumptions do not hold in practice.

Exchanges of information and movement of personnel are reciprocal among enterprises of one industry and between them and other enterprises, downstream and upstream in the industry. Moreover, because of the role played by technology, their results also affect the productivity of other factors of production (e.g., labor), so that they are "non-separable." As it is known, when externalities are reciprocal and non-separable, Pigovian compensations are not feasible (Nath, 1969).

Moreover, the skills and knowledge developed for one product can be applied to other products as well, which enhances the importance of the externalities mentioned above.

Such externalities are linked to the person-embodied character of knowledge, but externalities may also arise from product-embodied knowledge, through copying and imitation (reverse engineering). Although patent laws and other instruments for making knowledge proprietary limit the externalities from reverse engineering, they are not, as it is known, totally efficient.

The learning involved in the process of technological development implies also that the productive capacity of the resources used in this process increases over time. Such resources (especially manpower) can be used in the future not only by the enterprise which bears the cost of technological development, but also by others. The former cannot usually appropriate all the benefits deriving from such increased productivity. As a consequence, what are costs, in the short run for the individual enterprise are, in the long run and for the economy, the building up of productive assets.

The investment in technological capacity by an industry will probably induce similar investments by its suppliers, where analogous learning processes will probably occur. The combined effect of such learning processes, which are mutually supporting, leads to an increase in the general productive capacity of the economy.

However, learning provides external economies in the future. This time-dimension is of critical importance, as in the short run there may be offsetting diseconomies, especially because of the higher risks inherent to the early stages of learning.

The externalities discussed here can be seen as the result of differentiated learning processes by individual enterprises, and are an important element in leveling out the disparities of technological development within the economy.

Although informal exchanges of information, movement of personnel, and reverse engineering have an opportunity cost for the firm that originated the knowledge, for the society, such knowledge has no opportunity cost, as no further expenses are required for its transmission (Arrow, 1962a).

We argued above that the diversification of resources used in a strategy of technological self-reliance imply that it is less risky for a firm to pursue such a strategy if it operates in an environment where the technical division of labor has been considerably extended, so that the firm can have access to specialized knowledge. Because of the role played by person-embodied knowledge and informal exchange, physical and cultural proximity make this access considerably easier.[7]

For the same reasons, pursuing a strategy of self-reliance in such an environment is not only less risky, but cheaper for the firm. Not only will there be more possibilities of externalities, but resources will also be more effective. Such advantages apply, of course, more intensively to the more complex products — those requiring higher performance and reliability standards and those composed of a multiplicity of sub-systems.

The process we have been discussing is one of reciprocal stimulation: If the participants in such process are, in general, investing in their own scientific and technological capacity, the result, through this mutual stimulation, will probably be a higher general level of technological development. In other words, it probably results in what some authors have called "collective work" and still others "sinergy" — a total effect that is greater than the sum of its parts.[8]

It is probable, therefore, that external economies previously mentioned are subject to scale effects, increasing more than proportionately when the general investment in R & D increases. On the other hand, the same process of reciprocal stimulation applies to the use of foreign technology. For example, the use of foreign specifications by capital goods producers may lead their suppliers to use licensing too (or to be required to do so), and this, in turn, complicates the work of capital goods producers trying to develop designs locally, as the suppliers do not have an independent capacity to feed into this process. This implies that for self-reliance it is necessary not only to have an extended division of labor, but also a division of labor in which there is a widespread investment in science and technology. It is possible to envisage, in fact, a fairly developed industrial structure, where enterprises interact at the level of the production of commodities, but not at the level of the generation of the knowledge for producing such commodities, being all dependent on knowledge generated abroad. Such a "scenario" closely corresponds to the situation presented by the more industrialized developing countries, which have, through their process of industrialization, "redefined their terms of dependence" (Cardoso, 1973).

In fact, in many developing countries what we may call their "structural weakness" in terms of the supply of resources available to firms used for a strategy of self-reliance — the limited division of labor, the lack of trained personnel, the undevelopment of the technical and scientific information structure, etc. — has been compounded by a pattern of industrialization that placed an overall high degree of reliance on the use of imported technology, rendering more difficult the efforts of self-reliance of the national enterprises. We will return to this point in the next section.

State Intervention in an Open Economy

The reasons for state intervention previously discussed apply to any capitalist economy, although the peripheral countries' lower degree of capital accumulation and division of labor may accentuate the divergency between the logic of single capitals and capital in general, emphasizing the need for state intervention.

The preceding discussion abstracted from the influence of international conditions on the process of accumulation. However, one of the *diferentia specifica* of the peripheral countries is, besides their degree of capital accumulation and division of labor, their form of insertion in the international system.

As it is known, such a system has been increasingly affected by scientific and technological activities, politically and economically. At the same time, the internationalization of the results of such activities is often a necessary condition to make them economically viable, especially in the more R & D-intensive industries.

Therefore, it is necessary to set the issue of the intervention of the state in R & D in its international context. This context has three main features that will affect the action of the state in the science and technology area and create further reasons for its intervention: it is partially "open," in the sense of mobility of resources, competitive, and assymetric in the division of economic and political power.

The openness of the system provides the national economies with the possibility of substituting the science and technology of other countries for their own, importing the results of scientific and technological activities through movements of commodities, capitals, and knowledge itself (embodied in persons, publications, blue-prints, etc.).

This possibility seems, at first sight, highly attractive for those countries that have a poorly developed scientific and technological base and whose objectives of fast accumulation put a high opportunity cost on the resources allocated to the constitution of such a base. As we have mentioned, such advantages of being "latecomers" were highly emphasized by the literature of "development economics" of the early 1960s.

Subsequent research showed, however, that notwithstanding some advantages, such substitution first, was not perfect and, second, entailed some substantial costs not visible at first sight.

As for the first point, the concentration on the system's R & D activities in the central economies implied a "localization" of such activities around the specific economic and political problems of such economies, insofar as the growth of knowledge was influenced by the economic and social environment and not determined only by its own internal logic and that of the scientific community.

In the measure that the conditions of the other societies were different, there was a need to adapt technology to such conditions. Moreover, the diversity of conditions and the specificity of knowledge implied that there were opportunities of investment laying idle for lack of appropriate knowledge, as in the case of natural resources typical of the tropics.

Therefore, a minimal internal technical base was necessary to perform adaptations and the fulfillment of the possibilities of accumulation suggested the opportunity of extending such a base beyond the requirements of adaptation, which normally involve scientific and technological activities less complex than those required for innovations. Although not exclusively, such considerations applied especially to the peripheral countries. In fact, many of such countries, especially the more industrialized ones, have developed an adaptive capacity for imported technology, embodied in activities of detailed design of products and processes. However, as research concerned with "opening up" the R & D "black box" has shown, there is very often a discontinuity in the skills and resources needed for detailed design required for adaptation and the stages of R & D[9] where innovations are introduced, so that mastery of detailed design skills does not ensure a progress towards a capacity of introducing innovations, contrary to some more optimistic evolutionary hopes of the past. Although licensors of technology have a deep interest in teaching their licensees detailed design skills and production know-how, since their income usually depends upon the sales of the licensed products, it has been shown that the capacity for performing basic design and more complex scientific and technological activities is not transferred, so that the technical control is retained by the licensors and the licensing relationship is maintained and extended whenever there is technical progress, unless the licensees invest in their own technical capacity and use the licensing relationship as an auxiliary learning tool.

In other words, licensing allows the introduction of innovations (developed abroad) but does not create the capacity for introducing innovations. A policy of self-reliance, in contrast, leads to the development and diffusion of R & D skills.

Therefore, although the private firm's expenditures in licensing do lead to the creation of some productive assets for the economy under the form of increased technical capacity, the importance of such capacity is substantially less than that created by local R & D. For the same reasons, the externalities of information accruing from a licensing strategy are less important than those deriving from R & D expenditures.

However, in the context of "a mixed strategy," the skills developed through licensing can be used to reinforce those developed through local R & D, especially if there is a degree of complementarity between the products to which the strategies are being applied. In such case, the technical progress benefits can, to some extent, be realized without forsaking most of the benefits of licensing. In the long run, this may also allow a substitution of local technology for licensing.

The small amount of employment created due to the use of labor-saving techniques has been one of the most often cited problems deriving from the wholesale import of technology by the peripheral countries; it is also often indicated as a reason for state support of local technological development. However, to be effective, a policy of developing and using labor-intensive technologies seems to require that the process of capital accumulation be based on industries producing mass-consumption, low-cost goods that would be consumed by workers, and/or require that the workers hold enough political power to press the state for employment policies. Notwithstanding the potential welfare implications of a technological policy with employment objectives, the possibilities of such a policy being effectively implemented by the peripheral countries seem to be remote because one must bear in mind the pattern of growth of their economies which is based upon a highly skewed income distribution, the lack of political power of the workers, and the consequent representation of interests within and by the state in most of these countries and especially in the most industrialized.

In terms of employment, the main direct beneficiaries of the state's support of greater technological self-reliance will probably be the technico-professional groups. Their employment opportunities will be increased, as probably will be their incomes and their relative power within the state. In fact, such groups have often pioneered the support of such policies in the developing countries, within and outside the state.

In more general terms, if one were contemplating a change in the model of production, the mastery of technology would take on a much greater importance for the workers and other groups committed to such change, since such change, to be effective, would probably require new forms of organization of production (e.g. Bettelheim, 1968). However, such concern presently has probably little bearing on the adoption of policies of greater technological self-reliance in the capitalist developing countries we are discussing. Although it is an important theoretical and practical question, we shall not pursue it here in more detail.

The subtitution of an external scientific and technical basis for the internal basis can imply substantial costs to the local entrpreneurs too, leading them to press the state to act against such substitution.

The costs of substitution to the local entrepreneurs would be highest in the cases of industries already established where the substitution was made through the entry of foreign firms having a technical basis abroad that is stronger than the local capitalists', since, in such cases, the local firms would find their growth and survival under considerable threat, especially in the technology-intensive industries. The limitations placed by the Japanese State to the entry of foreign capital in strategic sectors and the attempts of the French to do the same in electronics can be understood in the light of such interests.

In cases where the substitution via foreign investment had a preemptive nature, i.e., in cases where the foreign firms practically set up the industry in the peripheral countries, as was often the case, for instance, in durable consumer goods, the grounds for conflict were, of course, much smaller, especially if the foreign investment opened up investment opportunities for the local capitalists. In fact, as the history of the industrialization of the peripheral countries shows, this type of substitution was actively encouraged by the majority of local entrepreneurs and national states.

The presence of technically strong foreign competitors in technology-intensive industries can provoke two distinct but not mutually exclusive reactions from their local competitors: They can either reinforce the local technical basis, investing in R & D and mobilizing the state to support their investment and to strengthen the scientific and technological system, or they can search for alternative sources of technology abroad. A distinctive trait of the situation in the peripheral countries has been the practically exclusive adoption of the second path, while in the core countries the solution has generally been a "mixed strategy," with higher expenditures for internal development.[10]

The strategy followed by the local enterprises in the peripheral countries is, to a considerable extent, a rational response to the deficiences of their local scientific and technological systems (due, in part, to their lack of investment in R & D in the past), to the pressures of a demand requiring products with foreign technology (producer and consumer goods), and, especially, the pressure of competition from foreign subsidiaries. In this sense, the latter contribute, even if unwittingly, to stunt the local scientific and technological development. Nonetheless, several studies[11] have shown that state policies of industrialization have played an important role in fostering such strategy, either directly by the lack of support to local scientific and technological activities and incentives to use foreign technology, or indirectly by contributing to the characteristics of the demand and competition prevailing in most industrial markets. Such policies of the states of the peripheral economies stand in contrast with the policies followed by the states of the core countries which, albeit selectively, have adopted a "package" of policies — ranging from financial grants for R & D to protection in the national markets and state purchases — which fosters the technological development of some key-industries, especially electronics, nuclear, aerospace, and machinery.

The import of "pure" technology through licensing and other forms of "technology transfer" could apparently solve some of the dilemmas of local capitalists and the state as regards the technical base, especially when the local base is poorly developed. Licensing enables the use of the external base at a relatively low nominal cost, reduces the risk and time by using products and processes already tested and, at the same time, supposedly retains the control of capital and investment opportunities in local hands.

However, the studies of "technology transfer" to the periphery have revealed several black spots on this rosy picture. In technology-intensive in-dustries especially, the costs of importing technology are quite often much higher than the nominal costs of licensing, being accrued by tied-in pur-chases of raw materials, components and parts. The risk is often not eliminated, sometimes because of different local conditions and sometimes because licensors use licensees to test new products or processes and, finally, the national control of decisions is substantially reduced, especially if the licensee does not possess an independent technical capacity.[12]

With regard to the last point, the studies have shown that licensing agreements normally contain restrictive clauses that limit the actions of their licensees, such as export prohibitions, free grant of improvements introduc-ed by the licensee to the licensor, etc. Moreover, licensing agreements are often used by licensors as a basis for the takeover of local firms (or, at least, controlling their strategic decisions) by receiving payment in equity form in-stead of money, as well as a means of testing the local markets prior to establishing a subsidiary, which will then compete with the erstwhile licensee, often with lethal effects for the latter.

No matter how much depreciated by the literature, and notwithstanding the regulations established by the states forbidding them, the restrictive clauses of licensing, especially those related to the control of innovations, to the control of access to markets, are inherent to the logic of the relationship. In the latter case, in fact, the prohibition to export is normally compensated by the monopoly given to the licensee in his local market, one of the main at-tractions of licensing. As for the technical control, the responsibility of its endurance rests with the licensee, who is not forbidden by the licensor to develop his own technical capacity which may, in the future, release him from the dependence of the licensor.

In fact, the available evidence suggests that there is a trade-off between licensing and local development of technology in the peripheral countries, at least in some industries. The licensee exchanges a relatively safe growth based on the internal market in the short run for the limitations of access to foreign markets and the possibility of being controlled and even ousted from the market in the longer run. The evaluation of such outcomes will depend upon several characteristics of the firm, such as the range of products it produces, its financial resources, and the time horizon it uses, as well as the constraints of the demand and competition. However, besides such economic factors, its choices will also be influenced by the value placed by its owners on political factors, such as being controlled by a foreign enterprise, a point which we shall return to later on.

Nonetheless, if a firm has an R & D capacity, by combining it with licensing, it can increase the benefits and reduce the disadvantages of R & D, especially by "spreading" its risks. At the same time, an R & D capacity allows the firm to obtain better conditions from licensing, to reduce its disadvantages, and to profit more from it in terms of learning and risk-reduction. In other words, the two strategies, when combined, interact, so that the effect of a mixed strategy is different from the summation of the effects of the two strategies separately.

From the social, as well as from the private, point of view, a mixed strategy with the two strategies combined in a complementary way is probably the optimal strategy. From the national point of view, the complementary strategy could be sought for products of different enterprises, so that a mixed strategy would be compatible with some enterprises being totally self-reliant, and others totally reliant, on licensing.

However, a mixed strategy may not be feasible under free market conditions, especially if there is an initial gap between the technology requirements and the local enterprises' experience, and such gap can be satisfied by licensing. In fact, the market mechanisms will stress the disadvantages of local R & D more in terms of risk and time than in its advantages of development of technical skills, which are, to a large extent, transacted outside the market mechanisms.

If the local markets are oligopolistic, and the strategies of the enterprises interdependent, this process is reinforced. If licensing gives one competitor an advantage, the others will follow. Under conditions of high concentration, the producers using licensing may even oppose state policies supporting local R & D if they can be used by new competitors.

As the economy grows, and technology requirements become more stringent, the market forces analyzed above will probably lead to an increasing reliance on licensing, with local technology being applied only to the simpler and older vintages of products.

Therefore, when there is an initial gap between technology requirements and the local enterprises' experience, and such requirements can be satisfied by licensing, a strategy of investment in R & D in conjunction with licensing would require state intervention. Given the risk of new technology, it is probable that an intervention via price mechanisms (e.g. lowering the cost of local R & D) will not be enough — a consensus may have to be established administratively and politically among the state, the producers, and the customers, as regards the development and use of local technology. Given the time-dimension involved, this consensus has to be structured around a long-term prospect — points to which we will return in the next section.

Comparing the situation between center and periphery, the main difference does not seem to be, as the literature sometimes seems to suggest, in the use of licensing by the enterprises in the former and local R & D in the latter, but mainly in the proportions used in the strategy mix. In the central countries, licensing is coupled to a strong capacity in R & D, reinforced by a continuous investment in such capacity, so that the firms are able to develop their own technology for new and complex products, as well as to maximize the benefits from a mixed strategy. In the developing countries, technological ability applies mainly to old or simple products, reflecting and reinforcing a predominant reliance on licensing for new products.

We must distinguish the rationale of local enterprise from that of foreign enterprise operating in the local market if the latter is a subsidiary of a multinational group. Even if the two are of similar size and producing the same range of products, the latter is part of a much greater block of capital with direct access to international financial, technical, and managerial resources, but which has a logic of accumulation subordinate to the worldwide accumulation of the group to which it belongs.

As regards foreign enterprises, the late Professor Johnson, a citizen above suspect of being biased against foreign direct investment in developing countries, aptly commented that:

> The corporation's concern in establishing branch operations in a particular developing economy is not to promote the development of that economy according to any political conception of what development is, but to make satisfactory profits for its management and shareholders. Its capacity to make profits derives essentially from its possession of productive knowledge, which includes management methods and marketing skills as well as production technology. It has no commercial interest in diffusing its knowledge to potential local competitors, nor has it any real interest in investing more than it has to in acquiring knowledge of local conditions and investigating ways of adapting its own productive knowledge to local factor-price ratios and market conditions. Its purpose is not to transform the economy by exploiting its potentialities — especially its human potentialities — for development, but to exploit the existing situation to its own profit by utilization of the knowledge it already possesses, at minimum cost of adaptation and adjustment to itself.

The corporation cannot be expected to invest in the development of new technologies appropriate to the typical developing country situation of scarcity of capital and abundance of unskilled, uneducated, illiterate labor, and in the mass training of blue collar, white collar, and especially executive local personnel. It has at its disposal an effective technology appropriate to the capital and skilled-labor-abundant circumstances of the developed countries. Hence, it will invest in technological research on the adaptation of its technology and in the development of local skills only to the extent that such investments hold forth a clear prospect of profit" (Johnson, 1970 b. p.26).

In fact, as a panel convened by the National Academy of Sciences of the United States concluded for several sectors examined: "(E)xcept for scaling down plant size to lower-volume markets, most international firms appear to have done little to adapt product design or production techniques to local conditions. They have made little effort to implant indigenous research and engineering capability that is innovative enough to adapt imported technologies to local needs and resources. Industrial technologies generally have been held on a close proprietary basis, especially if they involved advanced technologies with valuable competitive advantages" (NAS, 1973, p.20). Such conclusions are supported by other studies (e.g. Skinner, 1968), even allowing for some counter examples of adaptation (AID, 1972).

The reliance of developing countries' subsidiaries on their parent companies' technical and scientific resources is often explained in terms of economicity of use of resources: For the group, there is a clear rationale of concentrating their R & D and related services where such activities benefit from a highly developed level of division of labor and specialization, higher productivity of resources, and economies of scale. Recent research has shown that there is also a financial rationale in such concentration through the provision of an additional channel of flow of funds from the subsidiaries to the parent companies. The evidence available shows that, usually, the payments for technology from subsidiaries to parent companies are much higher (often by an order of magnitude of several times)[13] than between non-affiliated parties, and that such purchases of technology are very often tied to purchases of raw materials, parts and components at prices frequently much above their international market price (Vaitss 1974, Frenkel et al., 1978). Such payments for "technology" seem often to be used to by-pass local restrictions on profits remittances and, in the case of the developing countries, given the concentration mentioned above, they constitute a one-way flow of financial resources toward the advanced countries, which have a real-resources counterpart of doubtful value in productive terms.

Given the foreign exchange constraint that often bears upon the developing countries, the outflow of payments for "technology" is often the most "visible" part of the issue of technological dependence and, since foreign subsidiaries are often responsible for the majority of such payments, their role is most often discussed under such heading. But, as we saw above, they have an important effect upon the local scientific and technological system, leading their local competitors, suppliers, and customers to use foreign technology preferentially.

The process of accumulation of the peripheral countries often suffers from a foreign exchange constraint, which affects the growth of several sectors. In such conditions, a local technical capacity may be desired to alleviate such constraint, since it may result in:

1. Import substitution of finished products;
2. Import substitution of licensed technology;
3. Better conditions of bargaining for remaining technology imports;
4. Export expansion through the removal of barriers of technology;
5. Export expansion through new products.

As we saw above, the development of scientific and technological capability presents several features of the "infant industry" argument: minimum scale of activities, dynamic economies of scale and learning. However, since the foreign exchange constraint in the developing countries is often not reflected in the market prices — e.g., by keeping an overvalued exchange rate to benefit importers — the choice of strategies by the enterprises, according to market prices, will be biased and may not reflect the social opportunity cost of the resources used. Moreover, the foreign exchange effects of a growth of technical and scientific capacity will take time to be felt and, in the short run, the international system may provide alternatives, such as exports based on foreign subsidiaries, which have a side-effect of strengthening the resistance to greater technological self-reliance.

Nonetheless, the scope for solving the foreign exchange constraints of the peripheral countries based on exports of foreign subsidiaries seems limited and is subordinate to the interests of the parent companies of the latter, which may not coincide with those of the state and other groups represented by the state. If the capitalists of the periphery — presently facing protectionist policies of the advanced countries in their more "traditional," low-intensive technology markets (shoes, textiles, etc.) — wish to enter the high-growth markets of medium and high-technology, they will, by and large, have to rely on their own technology (because of the logic of licensing) and be supported by their states, last but not least, because of the support given by other national states to their competitors.

In fact, it deserves to be stressed that in the present conditions of many markets, where technology is a critical competitive asset and where most national states (especially the more developed ones) support the technological develoment of local enterprises, if a specific state does not provide such support to its national enterprises, the latter are automatically placed in a disadvantageous position in terms of international competition in such markets.

Self-reliance in some products may also be desired as an insurance against monopoly risks from abroad, as a means of reducing the uncertainty of the process of capital accumulation in the country, independently of the foreign exchange constraints previously discussed.

The state and national enterprises may wish to have an independent technological capacity for products strategic to their supply of licensing (or imports) being cut off by decision either of their licensors, or of their licensors' states [14]. From this point of view, self-reliance could, to some extent, override on opportunity cost grounds, considerations of cost of production, and efficiency.

Such motivation will depend on the number of alternative suppliers of technology and the degree of international alignment of the country with its main licensing parties. In this sense, on one hand, a close alignment of a peripheral country to some of the central countries may reduce its reasons for an autonomous technological capability; on the other, the lack of the latter may be an important dissuader of more independent polices in general, strengthening the hegemony of the central countries.

Our analysis suggests, then, that the pattern of development of the peripheral countries — their relative industrial and technical undevelopment, and their dependence on the center for consumption patterns, capital goods, finance, and technology (often internalized via foreign subsidiaries) — may lead to the constitution of a powerful bloc of interests, strongly represented within the state, which will probably attach a low priority to the development of a local scientific and technological capacity beyond that required for adaptation purposes, and may even oppose such policy altogether.

Nonetheless, in several of the peripheral countries, especially the more advanced, the state, mostly by its own initiative, has set up policies of incentives to local R & D performed by local enterprises. Such policies,[15] normally based on the reduction of the costs of R & D (via subsidized credits, fiscal incentives, etc.), combined to stricter regulations for the import of technology, in most cases are not powerful enough to counteract the market pressures discussed above. Nevertheless, they may have a restricted impact over a limited range of products, especially when combined to other measures of industrial policy and when the internal market has not been preempted by firms using foreign technology, especially multinational subsidiaries. Although structurally limited, such policies — if continued — will probably enlarge the economic and political basis of support through the linkage effects previously discussed.

Reproduction of the Social Conditions of Production

Internationally: Defense and Nationalism

One of the main characteristics of the state as a specialized institution in a complex is its "monopoly of the means of violence" for the purpose of upholding national sovereignty internationally, and keeping "law and order" internally. Such functions are linked not only to the possession of the means of violence but, to some extent, to the capacity to design and produce them.

As is well-known, the links between knowledge and weapons are intimate in the central countries: weapons systems are designed and produced based upon science and technology activities and the Second World War and, subsequently, the Cold War, have set the priorities of scientific and technological development.

War, however "is not merely a political act, but also a . . . continuation of political commerce, a carrying out of the same by other means"[16] and the denomination of "defense" given to the military expenditures of the central countries is, to say the least, misleading, especially when viewed from the periphery. Around such military expenditures in the central countries, a system of political and economic interests has been built that increasingly requires exports, of which an increasing share is being absorbed by the developing countries.

Such weapons exports, besides contributing to the arms race between the developing countries, contain the seed of a reaction — to free themselves from the arms dependency, the more advanced peripheral countries, such as Brazil, India, and South Africa, are setting up their own weapons industries, which will be truly independent only when they are able to design their own products. In fact, in such countries — as a replica of the central countries — the armed forces and the related industries are often among the main supporters of policies of greater technological self-reliance.

The internal political cohesion of capitalist societies and the role the state plays there, are based on the idea of the people as a unity — the body of citizens equal before the law, mutually responsible, and supposedly equally represented by the state. For such a concept to be effective in a world divided in nation-states, it is required that each nation-state be sovereign,[17] that the state will place the interests of its citizens above the interests of the citizens of other states, and that the main allegiance of the citizens of each state will be to their fellow citizens and to their state. Those are also the basic points of a nationalist ideology, as differentiated from nationalism as, simply, "a feeling of national consciousness — nationalism in a weak sense." (Kamenka, 1973, p. 15).

From this perspective, the control of local economic activities from abroad weakens the internal political and economic cohesion of the nation in a similar way to the weapons' dependence — it undermines the sovereignty of the state and it introduces a different category of citizens who have divided allegiances toward local and foreign citizens and toward the state.

Such control is especially threatening if it affects the activities around which the process of accumulation is structured. That is, in the case of scientific and technological activities, it will be resented more if the process of accumulation is structured around technology-intensive industries.

It is possible that such a threat is felt more strongly within the state because of its role in the political organization of capitalist society — which may explain why the initiative of nationalist policies, including those supporting science and technology, come from the state, especially where, as in the peripheral countries, there is a strong solidarity of interests between local and foreign capitalists. Nevertheless, even in such countries, nationalist policies still elicit wide support, especially among middle classes and, sometimes, among the workers.

Besides providing a strong motivation for state support to local design activities, a nationalist ideology may be a necessary condition to implement such policies.

First, nationalism provides the necessary basis for a long-term prospect and joint action by the state and the entrepreneurs, and it is the only ideological structure that allows such common action while preserving the separation of capitals.

Second, nationalism legitimizes the present sacrifices (unequally distributed) that such strategy requires, in the interest of future citizens (whose benefits are also unequally distributed). In countries where the wealth and income distribution is particularly uneven, as it is often the case with developing countries, this legitimizing function is likely to be especially important.

The less developed countries face a specially contradictory condition as regards nationalism. As an ideology, nationalism is likely to be especially strong among them, as a result of their colonial past, their present asymmetry regarding the more advanced countries, and their very skewed distribution of income and power. However, because of their undevelopment, they have to rely on the more advanced countries for finance, capital goods, and technology, which reinforce nationalism on one hand, and limits considerably the possibility of nationalist policies being applied in practice on the other. Where such reliance has been internalized through foreign subsidiaries, this has not only led local enterprises to rely on licensing, but has also established a common interest between the two groups, in terms of technological strategy and other policies, thus undermining the political and economic basis for nationalist policies. In fact, one of the most important distinguishng traits of the development pattern of the periphery in comparison with the center is the limited development in the former of a national bourgeoisie, defining itself not only in class terms, but also nationally (in contradistinction to other national bourgeoisies), a group which in most developing countries has not become the hegemonic partner in the power bloc[18].

Internally: Science and Technology as Power Resources

Science and technology are important power resources not only in the relationship between nations, but also within each nation.

At the level of the units of production, the separation between manual and intellectual labor and the parceling and hierarchization of tasks has concentrated the control of the process of production and, even more so, the control of the changes in such processes. However, the political implications of this type of technical progress are normally disguised under the cover of "technical imperatives," "scientific management," etc.[19]

Such "technical reason," prevailing in industry since the end of manufacture, has been progressively extended to other political relationships based upon the growing importance of science and technology and upon the idea of the "neutrality" of science, especially since the state increased its action to maintain the conditions of accumulation.

With the enlargement of its economic functions, the state, on one hand, has needed to think society scientifically increasing its scientific and technical personnel and setting up special institutions for this purpose (e.g., planning ministries). On the other hand, the state has used "technical reasons" to justify its actions. Therefore, scientific and technological knowledge have become, at the same time, instruments of intervention and political resources for justifying such intervention.[20]

The diffusion of a "technical rationality" can be an important instrument for the maintenance of the political cohesion, one of the prime concerns of the state. If the several classes and social groups accept that the solution of the economic and political problems is only a question of "social engineering," the focus of the debate turns upon the "technical quality" of the solutions. Thus, in the first place, the debate is "de-politicized," distracting the participants from its real political content and, in the second place, the number of participants in the debate is reduced, since in order to participate, "knowledge" is necessary. As access to knowledge is restricted and highly differentiated between classes, the final result is a substantial concentration of political power.[21]

The cohesion derived from the diffusion of a "technical rationality" is reinforced when local enterprises and the state succeed in developing scientific and technological activities with high public "visibility" and with high international prestige, which serve to legitimize further the existing regime. The priority given, internationaly, to "big science" and "big technology" activities seems to respond, in part, to this type of motivation and, at the same time, reinforces such priorities, orienting the states to such activities, which have a high political pay-off, nationally and internationally.

The diffusion of the "technical rationality" rests in good measure upon the education system, which, in most countries, is supported by the state and which receives a considerable share of the R & D expenditures.

However, at the same time, the education-and-research system also contributes to develop eritiques of the *status quo,* so that the state attitude to it is often ambivalent, especially in the more authoritarian regimes. For instance, in many Latin American countries the state, while supporting a local development of science and technology through some of its apparatuses, at the same time attacked the universities and research institutes through other apparatuses, often expelling scientists abroad and jeopardizing the activities of those remaining in the country.

Social cohesion, however, is not built solely around ideological features. In most countries, the state takes the responsibility for several activities that foster such cohesion, such as public health and sanitation, public transport, education, etc. Such activities increase the standard of life of the bulk of the population while, at the same time, providing entrepreneurs with a healthier, better educated labor force.

In the central countries, activities related to the population's health, especially pollution control, have recently increased their importance as users of research and development — often performed by private enterprises with state subsidies and in response to government regulations. It is worth stressing that the latter are, in their turn, a response to political pressures exerted by well-organized groups acting either through the state or through channels of the civil society.

Finally, when all other means of preserving a political regime fail, a state can resort to its *ultima ratio:* internal violence. Repressive apparatuses have, in recent years, increased their use of science-and-technology-based instruments, such as electronic devices and chemical weapons as gasses. Often, such means of violence can be used both for internal and foreign aggression purposes (Dixon, 1976).

The different ways through which the state keeps the political cohesion of the society, led to different demands to the science and technology system, and, consequently, to differentiated science and technology priorities and policies.

For instance, although the very low level of subsistence of the bulk of the population in the peripheral countries would suggest that a high priority be accorded to the activities related to, an increase in the standard of life and political co-optation (such as health, sanitation, etc.), the combination of authoritarian regimes and the low level of organization of the labor-force may put a low priority on such objectives. By the same token, such conditions may dispense the use of "scientific management" at plant level and, especially, a development of autonomous capability in such areas beyond the adaptation of imported techniques.

The same conditions, however, could put an emphasis on the development of science and technology for repressive purposes and for those activities which, by high public visibility and international prestige, could contribute to legitimize the political regime by giving to it a semblance of modernity and similarity with the central countries, without, at the same time, providing the local populations with the political participation (albeit limited) available in the core countries.

Conclusions

We saw above that there are some important reasons that could lead the states in the peripheral countries to support the development of a local scientific and technological system, even allowing for the possibility of imports from the center.

Given that different groups across society and through time are affected in different ways by the amount of resources and the type of products to which state support is directed, any normative statement about, first, the desirability of such support and, second, about which products such support should be directed, depends on the opinion of the commenter on the relative importance of such groups, now and in the future. In other words, the assessment of the desirability of state support of local R & D is inextricably bound to value judgments, especially as regards the desired prospect of the society one is considering.

Such values can be scrutinized on the basis of rational arguments, so that normative assessments of policy, although subjective, are not irrational. Although space precludes further discussion of the issue here, one should be wary of the pitfalls of wishful thinking. Science and technology policies are, by necessity, subordinate to more general policies and to the economic and political conditions that determine the latter and that set the specific priorities of scientific and technological development. For instance, although science and technology policy could be used for the purposes of employment creation and more equitable development in the peripheral countries, the preceding analysis suggests that the present economic and political conditions of many of such countries make this an unlikely result.

In fact, our analysis suggests that the pattern of economic development of the peripheral countries, and the political structures that express such development, tend to reduce the priority of a local development of science and technology in the peripheral countries, and to restrict the range of products and processes locally developed.

Nonetheless, the future history of such countries is not predetermined, nor is the present pattern void of contradictions. In many such countries, a local scientific and technological basis has been developed, often based on the common interests of factions of the state apparatus, technico-professional groups and, to a lesser extent, local capitalists. Such a combination of interests, if continued, may extend its range of action in the future, building up a local scientific and technological capacity with its own momentum. Nevertheless, such range of action seems to be structurally limited, unless considerable changes in the international system and in the peripheral countries' pattern of development supervene, when, science and technology may then perhaps fulfill their old promise of truly being agents of mankind's freedom.

Notes

1. For the world distribution of R & D personnel, see Annerstedt (1978); for patents, O'Brien (1974); and for scientific papers, Price (1967). Developing countries' share was 12 percent for manpower, 6 percent for patents, and below 3 and 7 percent respectively, for physics and chemistry.

2. Henceforth, "R & D" will be used as a shorthand for the whole range of scientific and technological activities — some of which are not included in the more formal definitions of R & D — which constitute the scientific and technological basis of the economy.

3. This approach is implicit in some of the most important papers on the subject, such as Nelsen (1959) and Arrow (1962), whose influence on the following analysis the reader will easily recognize. For a fuller discussion, see Erber (1977).

4. The possibility that, while the average return of the firm may be satisfactory, fluctuations in the average return may give rise to a series of losses or negative cash flows, causing bankruptcy.

5. Insurance markets that would "price" such risks are normally non-existent because of:
   i. the intrinsic difficulties of estimating the probabilities of the outcomes of R & D investment, which depend on the behavior of other economic agents, which the insurer cannot predict;
   ii. the difficulty of distinguishing between a "state of nature" and a decision of the insured — the "moral factor" problem (see Arrow, 1962);
   iii. because the base of the market (those who wish to be insured) would probably be small and the risk high, the premium of the insurance would be very high.

6. Even in advanced capitalist countries, at a time in which there was a relatively abundant supply of financial capital, the OECD has reported that credit for R & D projects was particularly scarce (OECD, 1963).

However, it should be noted that the evidence from the literature on the problem of lack of funds for more innovative (and risky) projects is far from conclusive and often contradictory. Grabowsky (1968) and Scherer (1965), for instance, studying the same industries (drugs and chemical petroleum) in the same country (United States) came to unexplained opposing conclusions as regards the effect of the availability of funds on R & D projects — the former claims that such availability was an important factor for the realization of such projects, while the latter found no evidence of this importance.

Other authors, such as Eads and Nelson (1971), have suggested that even small firms in the advanced countries can find appropriate risk capital in the market, if they show probable high rates of return, an assertion which is far from validated by the survey made by Bean *et al.* (1975) of the practice of venture capital firms (also in the United States), which "apparently plays little direct role in the early stages of technological innovation, but rather funds the production and distribution of relatively well-developed innovations" (p. 405), preferring also to concentrate operations in firms that are already growing, rather than on start-up projects.

However, it is important to stress that the discussion briefly mentioned above is related to a context in which there is a highly developed capital market. In developing countries, where the capital market is poorly developed, especially the risk capital market, it is much more probable that the funds for developing one's own technology have to come mainly from the firm's own resources or from its general credit on the existing capital market, except in the case of direct state intervention.

7. Geographical concentration of engineering activities is often observed — e.g., the Midlands in the U.K., São Paulo in Brazil. Freeman (1974) has suggested that a considerable advantage enjoyed by the American electronics industry *vis-à-vis* its European competitors lay in its much wider net of information and physical proximity.

8. "It is a question of not only expanding the individual productive forces but to create through the cooperation a new force that fucntions only as a collective force" — Marx (1963) p. 864. "Synergy . . . embraces a wide range of elements which ultimately result in a 2 + 2 = more than 4 effect. Synergy results from complementary activities or from the carry-over of managerial capabilities" — Weston (1970), p. 309.

9. See, for instance Judet and Perrin (1971) for process industries, and Erber (1977) for capital goods industries.

10. Comparing the ratio of R & D expenditures to imports of technology* policy we find that while, in the former the ratio is close to 12, in the latter it is close to 2. Notwithstanding the limitations of this type of comparison, the order of magnitude is unmistakable. Comparisons of strategies in individual industries, such as Erber (1977), and Frenkel et al. (1978) yield the same result.

11. See, for instance the studies of the Science and Technology Policy Instruments Project covering 10 developing countries (Argentina, Brazil, Colombia, Egypt, India, Mexico, South Korea, Peru, Venezuela, and the Republic of Macedonia in Yugoslavia), summarized in Sagasti (1978).

12. The evidence of such consequences of technology transfer is widely documented, for developing countries of all continents by UNCTAD (1974); by Vaitsos (1974) for Bolivia, Ecuador, Columbia, Peru, and Chile; by O'Donnel, Linck (1973), and Sercovitch (1974) for Argentina; by Wionczek (1973) for Mexico; by Fung and Cassiolato (1976), Erber (1977), Frenkel et al. (1978) for Brazil.

13. Zenoff (1970) estimates that average payments from affiliated foreign licensees of United States firms are seven and a half times higher than payments from non-affiliated licensees. Similar results have been obtained by Biato, et al. (1973), and Fung and Cassiolato (1976) for Brazil.

14. Examples of supply restrictions for political reasons have often been noted for capital goods (by the United States against France, Cuba and other socialist countries and, among the latter, by the Soviet Union against China), and probably extend to licensing, too. In fact, the flow of licensing from the West to the Eastern bloc has increased substantially after the thawing of the Cold War.

15. Sagasti (1978) provides a good summary of the policies applied in some of the main developing countries.

16. Clausewitz (1974), p. 119. See Rapoport's Introduction to Clausewitz's classical work for an analysis of the importance of a "neo-Clausewitzian" view in American international relations.

17. The concepts of "citizen" and of the "people-nation," date from the French Revolution, which, together with the Industrial Revolution (Hobsbawn's "dual revolution"), shaped the historical form of capitalism. "The source of all sovereignty," said the Declaration of the Rights of Man and Citizens, "resides essentially in the nation." And the nation, as a contemporary author put it, "recognises no interest on earth above its own, and accepts no law or authority other than its own" (see Hobsbawn, 1973, p. 80). Such principles, albeit in a mitigated form, still govern the concept of nation.

18. Hegemony implies ideological, as well as economic, dominance (see Gramsci, 1949 and Anderson, 1977). If a national bourgeosie were the hegemonic group in the developing nations, a "national project" in the sense of a relatively autonomous capitalist development would have been the dominant ideology there.

19. Writers of the last century had less qualms: Ure, celebrating the invention of Roberts'

---

*Between Japan (a central country often pointed out as a heavy user of imported technology) and Brazil (one of the main peripheral countries in terms of science and technology policy).

spinning machine commented "this invention confirms the great doctrine already propounded that when capital enlists science to her service, the refractory hand of labour will always be taught docility." Quoted in Dickson, 1974, p. 80.

20. A well-known analysis is Chomsky's (1971) study of the use of social scientists and psychologists in the conduct and justification of the Vietnam War, but Ezrahi's (1971) more general comment is also to the point: "the justification of decisions by reference to research or investigation committees has acquired in America a symbolic-ritualistic function similar to the medieval practice of linking important decisions to precedents and predictions from Holy Scripture" (p. 217).

21. Readers will recognize the influence of Marcuse (1961) and Habermas (1971) on the analysis above.

References

AID. 1972. "Applied Technology for International Development" — Office of Science and Technology, Washington, U.S.

Anderson, P. 1977. "The Antimonies of Antonio Gramsci," *New Left Review* n. 100, January 1977.

Annersted, J. 1978. "World R & D Survey" in V. Rittberger, "The New International Order and United Nations Conference Politics: Science and Technology as an Issue Area," UNITAR, N.Y., U.S.

Arrow, K. 1962. "Economic Welfare and Allocation of Resources for Innovation" in N. Rosenberg (ed.) "The Economics of Technological Change," Penguin Books, U.K.

Bean, A.; Schiffel, D.; and Mogee, M. 1975. "The Venture Capital Market and Technological Innovation." *Research Policy* Vol. 4 no. 4, October 1975.

Bettelheim, C. 1968. "Calcul Economique et Formes de Propriété," Librairie François Maspero, Paris, France.

Biato, F.; Guimarães, E.; and Figueiredo, M. 1973. "A Transferencia de Tecnologia no Brasil," IPEA/IPLAN Estudos Para o Planejamento n. 4, Brasilia, Brazil.

Chomsky, N. 1971. "American Power and the New Mandarins," Penguin Books

Clausewitz, C. von. 1974. "On War," Penguin Books, Harmondsworth, U.K.

Cooper, C. 1973. "Science, Technology and Production in the Underdeveloped Countries: An Introduction," *Journal of Development Studies,* October 1973.

Dickson, D. 1974. "Alternative Technology and the Politics of Technical Change," Fontana/Collins, U.K.

Dixon, B. 1976. "What is Science For?" Penguin Books, Hamondsworth, U.K.

Eads, G., and Nelson, R. 1971. "Government Support of Advanced Civilian Technology," *Public Policy,* Vol. 19, n. 3.

Erber, F. 1977. "Technological Development and State Intervention: A Study of the Brazilian Capital Goods Industry," D.Phil. Thesis, University of Sussex, U.K.

Ezrahi, Y. 1971. "The Political Resources of American Science" in B. Barnes (ed.), "The Sociology of Science," Penguin Books, Hammondsworth, U.K., 1972.

Frankel, et al. 1978. "Tecnologia e Competição na Indústria Farmacêutica Brasileira," FINEP mime.

Fung, S., and Cassiolato, J. 1976. "The International Transfer of Technology to Brazil through Technology Agreements — Characteristics of the Government Control System and the Commercial Transactions, mimeo, Center for Policy Alternatives, MIT, Mass., U.S.

Grabowski, H. 1968. "The Determinants of Industrial Research and Development: A Study of the Chemical, Drug and Petroleum Industries," *Journal of Political Economy,* vol. 76.

Gramsci, A. 1949. "Note sul Machiavelli, sulla Politca e sullo Stato Moderno," G. Einaudi Editore, 1974, Torino, Italy.

Habermas, J. 1971. "Toward a Rational Society," Heinemann Educational Books, London, U.K.

Hobsbawn, E. (1973. "The Age of Revolution," Sphere Books Ltd. London, U.K.

Johnson, H. 1970a. "The Efficiency and Welfare Implications of the International Corporation" in J. Dunning (ed.) "International Investment" Penguin Books, Harmondsworth, U.K.

_____ . 1970b. "The Multinational Corporation as a Development Agent," *Columbia Journal of World Business,* May/June 1970.

Judet, P. and Perrin, J. 1971. "A Propos du Transfert des Technologies pour un Programme Integré de Developpement Industriel." IREP, Grenoble, France.

Kamenka, E. 1973. "Political Nationalism — The Evolution of the Idea" in E. Kamenka (ed.) — "Nationalism — the Nature and Evolution of an Idea," Edward Arnold Ltd. London, U.K.

Marx, K. 1963. "Le Capital, Vol. I" in Marx: "Oeuvres," Vol. 1, Bibliotheque de la Pléiade, Paris, France.

NAS. 1973. National Academy of Sciences, "U.S. Internatinal Firms and R, D & E in Developing Countries," Washington, D.C., U.S.

Nath, S. 1969. "A Reappraisal of Welfare Economics," Routledge & Keegan Paul Ltd. London, U.K.

Nelson, R. 1959. "The Simple Economics of Basic Research" in N. Rosenberg (ed.), The Economics of Technological Change," Penguin Books, Harmondsworth, U.K.

O'Brien, P. 1974. "Developing Countries and the Patent System: an Economic Appraisal," *World Development,* September 1974.

O'Donnel, G., and Link, D. 1973. "Dependencia y Autonomia," Amor-rortu Editores, Argentina.

OECD. 1963. "Science, Economic Growth and Government Policies," Paris, France.

Pavitt, K., et al. 1974. "Government Policies towards Industrial Innovation," mimeo, University of Sussex, U.K.

Poulantzes, N. 1968. "Pouvoir Politique et Classes Sociales dans l'Etat Capitaliste," Librairie François Maspero, Paris, France.

_____ . 1978. "L'Etat, le Pouvoir, le Socialisme" Presses Universitaires de France, Paris, France.

Price, D. 1967. "Research on Research" in D. Arm (ed.) "Journeys in Science: Small Steps, Great Strides," University of New Mexico Press, U.S.

Sagasti, F. 1978. "Science and Technology for Development: Main Comparative Report of the STPI Project," IDRC, Ottawa, Canada.

Scherer, F. 1965. "Firm Size, Market Structure, Opportunity and the Output of Patented Inventions," *American Economic Review,* Dec. 1965.

Sercovitch, F. 1974. "Foreign Technology and Control in Argentinian Industry," D. Phil, Thesis, University of Sussex, U.K.

Skinner, W. 1968. "American Industry in Developing Economies — the Management of International Manufacturing," John Wiley & Sons, N.Y. U.S.

UNCTAD 1974. "The Role of the Patent System in the Transfer of Technology to Developing Countries," N.Y.

Vaitsos, C. 1974. "Intercountry Income Distribution and Transnational Enterprises." Oxford University Press, London, U.K.

Weston, F. 1970. "Conglomerate Firms" in B. Yamey (ed.) "Economics of Industrial Structure," Penguin Books, Harmondsworth, U.K.

Wionczek, M. 1973. "La Transferencia de Tecnologia en el Marco de la Industrializacion Mexicana" in M. Wionczek (ed.) "Comercio de Tecnologia y Subdesarrollo Economico," UNAM, Mexico.

Zenoff, D. 1970. "Licensing as a Means of Penetrating in Foreign Markets," *Idea,* vol. 14 n. 2.

PART THREE

Technology and the International System

7

Technology Transfer and North/South Relations

Some Current Issues

FRANCES STEWART

Introduction

Questions related to technology are now central to much North/South dialogue.[1] But this is of fairly recent origin. In the early years after independence, there was little conflict between North and South on technology: most developing countries wished to maximize the technology inflow, and, apart from general incentives, pursued a laissez-faire policy toward technology. It was only when the results of such a policy became apparent, that countries began to follow more interventionist and aggressive policies, and technology became an important issue in North/South relations. This essay traces the developments that led to the current debates, and considers how technology influences international economic relations today.

The drive to industrialize was the prime aim of economic policy for many developing countries in the years immediately following independence. There were a number of underlying motives. In the first place, the developed countries were also the industrialized countries, and industrialization was thus regarded as synonymous with development. Secondly, excessive reliance on primary production in international trade was believed to be responsible for dangerous fluctuations in foreign exchange earnings, and also, according to Prebisch, for long-run deterioration in terms of trade, which would unavoidably move against primary products and in favor of industrial products. Third, the unprecedented rapid growth in population led to a parallel growth in the labor force. None of the traditional sectors could provide a sufficient number of productive employment opportunities, so governments looked to employment expansion in the industrial sector to absorb the job seekers.

During this period, technology policy was designed to support industrialization policy. Almost without exception, the developing countries lacked even the elements of a local technological capacity, on which industrialization might be based. Thus, rapid industrialization required the import of technology from the developed countries. The form of industrialization determined the nature of technology to be used. Most countries adopted an import substitution (IS) strategy, as being the only possibility in the initial stages of industrialization. IS offered protected markets for the incipient, relatively inefficient industries. But it was not just a matter of import substitution − the substitution of home-produced goods for previously imported goods − but *Import Reproduction* (IR), or the exact replication of previously imported goods, that was the general strategy: home produced Coca-Cola replaced the imported vintage; home-produced Palm Olive (or Tide or − at later stages − Renaults) replaced the imported. Such complete reproduction inevitably required the use of the same technology as that previously used for producing the imports − i.e., the use of the imported technology. Thus, a near exclusive reliance on imported technology was the result of a weak indigenous capacity, the requirements imposed by speedy industrialization, and the import reproduction policies followed. The international companies, which produced the goods previously imported, strongly supported the IR strategy, since they now replaced their sale of finished goods by supplies of finance and of management, and by the sale of technology and parts.

In their desire to speed up industrialization, most developing countries introduced general incentives to technology transfer from abroad. Apart from tax concessions, these included guarantees about convertibility and exchange rates, tariff protection and markets, and tariff remittance on inputs. Tax incentives include tax holidays, investment allowances and investment grants, accelerated depreciation, loss carry-over provisions, exemption from taxes such as stamp taxes, sales taxes, and so on. The most widely given direct tax concession is a holiday for five or more years. A survey of projects undertaken by firms from OECD countries in developing countries found that nearly half received tax holidays, generally of between five and seven years.[2] All 13 countries surveyed by Lent provided tax holidays ranging from two to 25 years. Other significant incentives are generous tax treatment of investment and special tariff arrangements. In one sample[3], more than a quarter of the cases received accelerated depreciation allowances, while in over half the cases tariff concessions were provided on imported parts and raw materials.

The policies adopted were undoubtedly successful in leading to a large inflow of foreign technology and a rapid rate of industrialization in most developing countries. Between 1960 and 1977, the share of industry in GDP rose from 17 to 25 percent among low income developing countries, and from 32 to 36 percent among middle-income countries. For middle-income countries, the share in 1977 was only one percentage point less than that for industrial countries.[4] During this period, developing countries accounted for an increasing proportion of world manufacturing value added (and also world trade in manufactured goods); in 1960 the developing countries' proportion of world value added in manufacturing was 6.9 percent; in 1977 it was 9 percent.[5] Direct foreign investment to developing countries amounted to $7.6 billion in 1976, of which $6.1 billion was to developing countries other than OPEC and tax havens.[6] This amounts to around 10 percent of industrial investment in developing countries (and a higher proportion of manufacturing investment). But there are considerable disparities between countries. For example, in 1977, Brazil alone accounted for 18 percent of the total flow. The figures for the flow of foreign direct investment cover only a proportion of the total transfer of technology from abroad. Technology is also transferred to locally owned firms — through technology contracts and machinery sales, for example.[7] Moreover, much direct foreign investment does not consist in financial flows, but in managerial and technology transfers, while finance is provided locally. Until recently, almost all investment in modern industry relied on technology transferred from the industrialized countries.

Inappropriate Technology

The policies were successful in their main objective — speeding up industrialization and the growth of GNP — but industrialization and growth did not live up to expectations in many respects, while the unregulated transfer of technology created new problems that have come to be central to international relations. A major objective of the industrialization policies had been to create productive jobs for the rapidly growing population and to absorb the large number of people who were unemployed or underemployed, working to low capacity and at low productivity. But the industrial technology — designed for the capital-rich industrial countries — was labor saving. Growth in employment lagged behind growth in industrial output, as Table 7.1 indicates for selected countries. Where the industrial sector formed

TABLE 7.1   Changes in Output and Employment in Percent Per Annum

	Output	Employment
Philippines	5.4	2.0
India	5.4	2.6
Korea (Rep. of)	21.1	12.7
Kenya	8.2	5.6
Peru	6.9	4.1
Brazil	11.3	4.9 (only Sao Paulo area)

Source: Jacobsson (1979)

only a small proportion of the labor force, very large rates of expansion in employment would have been required to absorb the additional workers. For example, in India the manufacturing labor force forms 9.5 percent of the total labor force. With a yearly increase in the size of the work force of 2.1 percent, an annual increase in manufacturing employment of over 20 percent would be required to absorb all the extra entrants into the work force. Yet, the actual increase in manufacturing employment was 2.6 percent per annum.[8]

The high capital-cost of each job created by the new technology was far in excess of the level of investment most countries could afford for each member of their work force. The technology designed for use in the industrialized countries used the sort of resources these rich countries could afford: with incomes per head ten to 20 times those of the poor countries, there were similar disparities in savings. Thus, while many developing countries' savings amounted to under $500 per head of the work force, typical investment costs of the industrialized countries were in excess of $5,000 per head. The transfer of this technology meant that the poor countries had to concentrate all their savings on the minority employed in the modern sector, to the neglect of the majority of their work force. It was not only a question of investment resources — technological improvements, social, and economic infrastructure all had to be concentrated on serving this sector, if its productivity were to be maintained. Thus, major disparities emerged between the resources used by the modern sector and those of the rest of the economy. The result was corresponding disparities in productivity and in incomes between the sectors. The problems of growing maldistribution of income and of extreme poverty in the rural and urban informal sectors was in large part due to the nature of the technology transferred.

Technology transferred from industrial countries is designed to produce the sort of products consumed in advanced countries — standardized, packaged, advertised, and designed for consumers with incomes per head of $2,000 or more. The products that came with the technology were well suited to the growing elites in developing countries, which had similar income levels to those for whom the technology was initially designed — but were generally inappropriate, being far too high-income — for the majority of poor consumers in developing countries.

The technology transferred was thus inappropriate in many respects — both as to the nature of the methods of production (capital-intensive and labor saving, large scale, and skill-intensive), and as to the nature of the products the technology produced. The incentive system that encouraged the massive transfer exacerbated the situation, since the incentives were largely related to investment expenditure, and hence encouraged capital-intensity, as did the tariff-protection system. The result tended to be a dualistic pattern of production — with a small modern sector that absorbed most of the countries' resources, to the neglect of the rest of the economy.

Cost of Imported Technology

Not only did the imported technology prove inappropriate, it was also extremely expensive. UNCTAD[9] put the overt cost of technology transfer at 5 percent of non-oil exports in 1968 — the order of magnitude is similar to the total inflow of private foreign investment. These are direct costs: but much of the cost of imported technology does not appear as overt payments (royalties, for example), but is disguised in the form of inflated prices for inputs (such as managerial fees, parts, etc.), or reduced prices for exports. The very large proportion of international trade that is intra-firm (i.e., between two parts of the same firm) permitted such practices.[10] Detailed studies of a number of industries[11] have indicated the extent and magnitude of such disguised payments. For example, in Colombia the absolute amount of overpricing of inputs of foreign firms amounted to six times the level of royalties and 24 times the level of profits.[12]

It is impossible to get reliable aggregate cost estimates because of the hidden nature of much of the costs. Moreover, technology usually comes as part of a package incorporating management, machinery, marketing, and sometimes finance even if the "hidden" element is correctly identified, it is not necessarily due to the technology transfer, but may be the price of other elements in the package. These difficulties arise, in large part, because of the fundamentally non-competitive nature of much of the technology market. Knowledge, which is essentially what technology is, may be expensive to produce, but once produced, costs of spreading it are fairly low. There is a very large gap between the costs of production and the costs of communication. In a competitive market, the price of technology would be lowered to the costs of communication, but in such a situation the costs of producing knowledge would not be covered. Hence there would be underinvestment in the production of knowledge. Thus, so long as the production and communication of knowledge remains primarily a private activity, motivated by the desire for profit, a non-competitive element has to enter the market. In the world market for technology today, there are two aspects to this non-competitive element: one, a legal system that protects the owners of knowledge (i.e., the system protecting patents and trademarks); two, an oligopolistic industrial structure that has developed, partly as a result of the legal protection, but also for more powerful reasons, which permits the owners of technology to levy charges in excess of communication costs.

Thus, it is of the essence for a system of technology transfer based on the sale of technology for profit, that the market should be noncompetitive and the prices charged should be excessive, if judged by the standards of a "competitive" price. Viewed from the point of view of the technology suppliers, this is regarded as a perfectly legitimate way of securing the appropriate returns on investments in the production of knowledge. But, from the point of view of the technology buyers, it seems that excessive prices are being charged. Thus, dissatisfaction and acrimony, as between buyers and sellers, is a natural development. Moreover, given the nature of the market, it really does not make sense to ask what the *right* price is. Essentially, it is a matter of bargaining because of the gap between the minimum for which the seller would part with the technology once known (roughly the communication costs), and the cost of reproducing the knowledge, which sets an upper limit on what the buyer is prepared to pay. When negotiations are conducted between firms, they are similar to other commercial transactions — some settlement is normally reached without a great deal of acrimony. In the protected environment of most developing countries, the buyer can normally retrieve any costs he pays in the form of high prices for the products he sells and, thus, there is little incentive to bargain roughly, or to complain about any agreement reached. Moreover, the buyer is often closely associated with the seller — indeed in the case of multinational investment, they are the same entity. There are close ties in other arrangements — e.g., joint ventures. It is only when negotiations are subject to national scrutiny that they become a matter of international relations and international dispute.

Over the last 20 years, governments of developing countries have taken an increasing interest in the technology transactions of firms importing technology. This is a response to the factors just described — the high overt costs, identification of substantial hidden costs, and the major potential for reducing costs by bargaining. National efforts are required to improve bargaining because individual firms are neither sufficiently strong, nor sufficiently motivated to bargain toughly. UNCTAD has contributed further to bringing technology transactions into international debate by trying to raise the question from a national to a North/South level, believing that, just as nations are in a better position to bargain than individual firms, so coordinated bargaining among developing countries will be more effective than national efforts.

Independence

The most important motive underlying the actions of many developing countries during the past half century has been the desire for independence. Those countries which had achieved political independence soon recognized that this did not bring with it economic independence. The drive to industrialize and, in many countries, to nationalize productive assets, was the next stage in the attempt to achieve independence. But neither industrialization, nor nationalization proved sufficient to secure local control over local activities. Complete independence is an impossible aim in an interdependent world, but this does not mean that complete dependence is unavoidable. A major factor in continued dependence is technological dependence. Technological dependence forces countries to leave the main decisions to their technology suppliers. It is an intrinsic part of direct foreign investment that control should remain with the foreign firm. But, close examination of the details of technology contracts between "independent" buyers has revealed an almost identical amount of control involved in arms length transactions.[13] Technological dependence is also responsible for dependence in more amorphous — but also more important — ways. Both consumption patterns and production methods are taken from the technology suppliers. This leads to cultural and psychological dependence, quite apart from its economic consequences.

New Objectives

As the consequences of the unregulated inflow of technology came to be realized, new objectives evolved. These objectives concern the characteristics of the imported technology, its costs, and its effects on local decision making and local technological developments. Developing countries differ substantially in the vigor with which they support the new objectives. While there is considerable concern about the limited employment provided by much industrial technology and the problem of unemployment and underemployment, the idea of "appropriate technology" has gained more support among developed countries than among the spokesmen of developing countries. There is a suspicion that appropriate or intermediate technology is inferior technology — a new device promulgated by developed countries to maintain developing countries in a state of underdevelopment. The situation is exacerbated by the fact that the developed countries have sponsored a number of appropriate technology institutions.[14] Properly interpreted, appropriate technology involves a major shift in resources and technology away from the elitist modern sectors and towards the mass of working people in developing countries — both with respect to production and to consumption technology. Suspicion of appropriate technology may stem in part from this fact: the potential gainers are not fully represented among developing countries' spokesmen. The modern technology in use in developing countries has created strong interests in its perpetuation. There are, nonetheless, a large number of developing countries' scientists, technologists, and economists working to promote an effective appropriate technology.[15] But at present, this is not a major issue of North/South relations.

On the surface, at least, there is much more universal agreement among developing countries about the need to reduce the costs and diminish the restrictions imposed by technology transfer. A number of countries have introduced active regulatory policies. The Andean Pact led the way; but many others[16], including Mexico and India, have introduced quite complex regulatory provisions. In general, the policies favor the *unpackaging* or *unbundling* of technology — i.e., moving away from the package presented by direct foreign investment and towards purchasing the various elements separately. There is growing evidence that reducing the bundling tends to increase the bargaining power of the purchaser, increase the learning effects of the technology transfer, and reduce the loss in control. The extent of bundling is a continuum: At one extreme, direct foreign investment represents complete bundling, and at the other, there is separate purchase of each item in the bundle. In between there are varying possibilities, including joint ventures, turnkey factories, etc. Developing countries' ability to unbundle depends on the sophistication of the technology involved and their own technological and managerial capacity. The more sophisticated countries are now in a position to unbundle a very large proportion of their technology. In addition to unbundling, policies regulate the terms of technology and prohibit restrictive clauses. On paper, some of the policies are fairly rigorous and have apparently resulted in large gains. For example, in Mexico between 1973 and 1978, the Registry examined 4,600 agreements and rejected 35 percent for excessive payments or restrictive clauses.[17] In Colombia between 1967 and 1971, the *Comité de Regalias* evaluated 395 contracts, modifying 3,334 and rejecting 61. In the process of renegotiation, royalties were reduced by 40 percent. From 1970–71 the *Comité* reduced tie-in clauses by 90 percent, eliminated export restrictive clauses, eliminated 80 percent of the minimum royalty payments, and established maximum percentage royalty rates.[18] The potential gains from active regulation appear high. But there are also costs: These are partly administrative, and partly concern a possible reduction in the inflow of technology resulting from the regulations.

Whether such a reduction occurs or not depends on whether the previous price was at or above the "competitive" price. Assuming it embodied some monopoly rents, then a reduction should be obtainable without loss of inflow. Limited evidence suggests this is the case. Figure 7.1 shows no noticeable reduction in the inflow of investment in Andean Pact countries following Decision 24, which introduced a number of regulatory policies.

An important question that needs to be answered is the extent to which these regulatory policies really are effective. Two factors may reduce their effects: first, the way in which foreign technology suppliers may manipulate the developing country decision makers so as to reduce the real burden of the regulations. Second, the extent to which the suppliers are in a position to replace *de jure* by *de facto* controls. Sagasti believes that both have occurred. "A potential licensee who needs technology is in fact at the mercy of the licensor who has the technology. No amount of government intervention can restrict the pressures imposed by a licensor, and most licensees submit voluntarily to the demands, although the formal contract may not reflect any controls."[19]

So far, effective bargaining has occurred only at the national level. No substantial international machinery has been introduced, although there has been much discussion of the need for it. It seems likely that the sort of problems encountered at the national level would be magnified at the international level. The experience of the Andean Pact illustrates this.[20] Although in theory each country adopted the same regulations, the actual execution of the policies has been nominal in some countries and effective in others. One country — Chile — has actually left the Pact. The major problem is at the level of political economy. The "national" interest may dictate tough policies, but nations are composed of non-homogenous groups whose interests may be variously affected by regulations on foreign technology. Where external interests have been effectively internalized, there is a Trojan horse already *in situ* to undermine any regulations.

The most genuine agreement among developed and developing countries concerns the desirability of building up developing countries' technological capacity; for the developing countries, this represents a way of increasing their overall independence, of improving their bargaining capacity and thus reducing the costs of importing technology, and, in the long run, offers the only serious ways to develop an alternative technology. For the developed countries, the policy offers no real threat, at least in the short run, particularly if it is based on "positive" measures (i.e., to local technological developments) rather than "negative" measures (i.e., restrictions on the import of foreign technology).

The major, unsolved problem is how to achieve genuine development of local technological capacity. Expenditure alone is not enough, as shown by

FIGURE 7.1   U.S. Direct Investment in Andean Manufacturing
in Millions of U.S. $

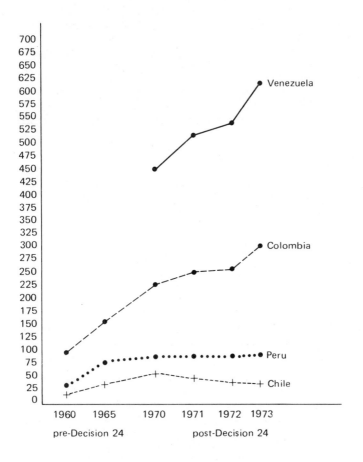

Sources: U.S. *Statistical Abstract of the United States 1972, 1973, 1974, 1975* (Washington, D.C.: Government Printing Office).
Source: Mytelka (1977).
University of Aston, Birmingham, 1979.

the many research laboratories in developing countries that produce very marginal results. According to Ranis, "the less developed world is strewn with scientific institutes and other expensive white elephants which contribute neither to science nor to technology."[21] It is known that it is necessary to secure close ties between the scientific and technological resources and the productive system. Yet, the level of research and development conducted by firms in developing countries is notoriously low. In consumption goods, particularly, foreign technology and foreign brand names guarantee marketability of the product. Hence, it is risky to use local technology.

Two alternative strategies for developing countries' technological development are possible. One involves radical divorce from the use of foreign technology, changing local consumption patterns toward local goods, and severely restricting the entry of foreign technology. The other involves using foreign technology as the main basis for local development, while local resources give an appropriate twist to the imported technology that increases the efficiency of the foreign technology. The former strategy is close to that of Mao's China; the latter is the strategy of the open economies such as South Korea. The first strategy requires a much more radical break with the past — and may involve substantial costs in the short to medium term. A strategy which continues to rely primarily on foreign technology as the basis for local innovations inevitably leads to a continued locking in to the patterns of development of the advanced countries. With some local adaptions, both production and consumption patterns are likely to follow developments in the industrial countries. A more radical delinking of technology is required for more autonomous developments.

The question of appropriate policies in developing countries toward the import of advanced country technology is an unsettled one. It depends, in part, on a country's development strategy. A strategy that involves inequality and elitist consumption patterns is best served by foreign technology. An outward-oriented development strategy, in which trade in manufacturer with the North provides the main impetus, also almost certainly requires heavy reliance on sophisticated advanced country technology. Economies that are inward looking, or look to trade with similar developing countries for their markets, would benefit from a delinking in technological terms. A second issue is the effect of foreign technology on local technological developments: While unregulated technology inflow may inhibit or swamp local developments, some foreign technology may be essential to stimulate and assist local development. Both Japan and the Soviet Union pursued policies of *selective* imports of technology. Japan used foreign technology to copy, learn, and adapt. In the 1930s, the slogan was "the first machine by import, the second by domestic production."[22] However, some restrictions on the free import of advanced country technology appears necessary to stimulate local developments. As Sagasti puts it: "Although technology imports do not always hamper the growth of domestic science and technology capabilities the wholesale importation of technology without efforts to screen, control and absorb it usually stunts the growth of domestic science and technology capabilities."[23]

The appropriate policies also depend on the stage of development — and the technological capacity — of the country. The very poor and undeveloped economies have little option but to import technology; but for them, the possibility of importing technology from other Third World countries is often an alternative to the import of technology from the North. It has the advantage of being generally more appropriate, with simpler products, and smaller scale and less capital-intensive technology.[24]

International Relations and Technology: Some Conclusions

The main areas of debate in the North/South dialogue are clear: They relate to the characteristics of the technology, its price, and the significance and implications of technological dependence. Analysis of these questions suggests that there are not homogeneous groups within either North or South who have particular and unique objectives in relation to each issue, but rather different interest groups — particular within the South — with diverse objectives. Consequently, despite the fact that the technology issue is normally debated on North/South lines, these may be the *wrong* lines, leading to a concealment of many of the issues, and a lack of decisiveness in results. For example, on the question of characteristics, there is divergence of interests between those within the modern sector in developing countries whose interests, both as producers and consumers, broadly coincide with the sellers of the technology in the North, and the remainder of the economy. On costs, while the sellers obviously gain by a high price, in a protected environment the buyers do not lose; the losers are the consumers of the products and the "national" interest. On the question of technological independence, who gains and who loses depends on how the policy is interpreted. If interpreted purely as a promotional/incentive policy to provide finance for local developments, there are few specific losers. But if interpreted as involving rigorous selectivity toward foreign technology, then foreign suppliers lose.

When it comes to international negotiations and national action, the reality and strength of the negotiating stance depends largely on whose interests are being represented. Appropriate technology tends to go by default because the underprivileged who would gain are not represented. Even on the question of costs, the fight put up by the developing countries is often more apparent than real: This is illustrated by the tax system. The generous tax incentive toward foreign technology continues in many countries, despite the fact that these represent a simple and obvious way in which the terms of technology transfer could be substantially improved.[25]

The imbalance in world expenditure on research and development is well known. The most recent estimates suggest that 97 percent of world research and development is spent in developed countries, and only 3 percent in developing countries.[26] Six industrialized nations employ nearly 70 percent of the world's research and development manpower, and spend nearly 85 percent of research and development funds. Only 6 percent of an estimated 3 1/2 million patents issued in 1972 were granted by developing countries, and less than one third of these were owned by developing country nations[27]. The world imbalance is not simply a question of aggregate resources, but of the *use* of those resources within both rich and poor nations. Within rich nations, a high proportion is devoted to defense and to aerospace; within poor nations, most of the technological resources are concentrated on duplicating efforts of the rich nations, on basic research, and on producing products for the elites. If we were to divide the world by income group and activity and allocate technological resources accordingly, we would find a much worse imbalance: While well over half of mankind survive on incomes arising from low-productivity activities outside the modern sector, the resources devoted to improving technologies in such traditional activities are miniscule. It is this imbalance which should be the main topic for international discussions of technology; but because of the nature of international debate — conducted by governments along North/South lines — the needs of the really deprived tend to be neglected, while discussion is concentrated on the relative positions of the elites of the world.

As a first step, what is needed is the development of a *Technology Matrix* for technological activity (Table 7.2). Filling out this matrix — in a very rough manner — indicates the kind of imbalance described above. The

TABLE 7.2 Technology Matrix

Who Benefits	Who Carries Out for Technology Activity:	Developed Countries: Public Sector	Developed Countries: Private Sector	Developing Countries: Public Sector	Developing Countries: Private Sector MNCs	Developing Countries: Private Sector Other: Large	Developing Countries: Private Sector Other: Small
Developed Countries		**	**				
Developing Countries							
Modern Sector: Industrialists Bureaucrats Workers Large Farmers		*	*	**	**	**	*
Traditional/Informal Sector: Urban Rural				*			**

Note: Two stars indicate major beneficiaries; one star indicates some benefit.

underprivileged — those engaged in small-scale activities, in the cities and the countryside, in agriculture services and industry — receive no benefit at all, or only small benefit from technological activity in the developed countries and from most activity in poor countries. It had been shown that even public sector activity[28] in developing countries rarely confers major benefits on the underprivileged. The underprivileged are the main beneficiaries of the technology they develop themselves — but in most countries, there is extremely little activity of this kind.

The benefits conferred by technological activity are likely to vary from country to country, according to their political and economic strategies, as well as their "explicit" technology policies.[29] A technology matrix needs to be worked out for each country. The effects of technology policy may be viewed in terms of how it affects the matrix of beneficiaries. This could provide a powerful tool for assessing technology policy both at the national and international level.

Notes

1. As shown, for example, by the attention given to it in the Group of 77's demands in the new International Economic Order and the many International Conferences on the subject — e.g., the United Nations Conference on Science and Technology for Development (Vienna 1979).

2. Reuber (1973).

3. Reuber (1973). See Lent (1967) and the Third Report of the meeting of U.N. Experts on Tax Treaties between Developed and Developing countries (U.N. 1972) for a description of tax incentives in developing countries.

4. *World Development Indicators,* World Bank, 1979. It is significant that during the same period there was a marked fall in the share of GDP accounted for by industry in "industrialized" countries, from 40 percent in 1960 to 37 percent in 1977.

5. *Industrialization for the Year 2000: New Dimensions,* UNIDO 1979.

6. *Ibid.*

7. See C. Cooper and F. Sercovich (1971).

8. Methodology and empirical estimates for a number of countries are contained in Stewart (1978), pp. 50–54. These figures are taken from Jacobsson (1979).

9. UNCTAD (1975a).

10. One-third of U.S. exports were estimated to be intra-firm in 1970 — see Lall (1973).

11. See e.g., Vaitsos (1974), Lall (1973).

12. Vaitsos (1971).

13. Clauses determining the nature and source of inputs, the quality and quantity of the output, the markets, and limits on local research development are quite common. See Sercovich (1974) and UNCTAD (1975a).

14. E.g., the U.S. Appropriate Technology International; the U.K. Intermediate Technology Development Group.

15. See the report of the Consultations undertaken for the proposed International Mechanism for Appropriate Technology (IMAT), Report to Dutch Government, 1978.

16. See UNIDO (1978).

17. UNIDO (1978).
18. Vaitsos (1971).
19. Sagasti (1978), p. 19.
20. See Mytelka (1977).
21. See Beranek and Ranis (1978).
22. UNCTAD (1978).
23. Sagasti (1978).
24. See e.g., Wells (1978), (1979).
25. See Stewart (forthcoming).
26. Annerstadt (1978).
27. See UNCTAD (1975).
28. See Warman (1978), Vallianatos (1976), and Beranek and Ranis (1978).

29. See Sagasti (1978), who distinguishes between implicit and explicit policy instruments, and argues that the implicit (including most significantly overall economic strategy) are of the greater significance. See also Ranis (1979).

References

Annerstadt, J. 1978. "Technological Dependence: A Permanent Phenomenon of World Inequality?" (mimeo, Institute of Social and Economic Planning, Roskilde University Center)

Beranek, W. and Ranis, G. 1978. *Science and Technology and Economic Development* (Praeger)

Cooper, C. and Sercovich, F. 1971. "The Channels and Mechanisms for the Transfer of Technology from Developed to Developing Countries" (UNCTAD: TD/B/AC.11/5).

Jacobsson, S. 1979. "Technical Change, Employment and Distribution in LDCs," The Lund Letter on Science Technology and Basic Human Needs, No. 13.

Lall, S. 1973. "Transfer Pricing by Multinational Firms," *Oxford Bulletin of Economics and Statistics,* 35.

Lent, G.E. 1967. "Tax Incentives for Developing Countries," *I.M.F. Staff Papers,* 14.

Mytelka, L. 1977. "Regulating Direct Foreign Investment and Technology Transfer in the Andean Group," *Journal of Peace Research,* XIV.

_____ . 1978. "Licensing and Technological Dependence in the Andean Group," *World Development,* 6.

Ranis, G. 1979. "Science and Technology in Development Planning: The International Dimension," Symposium in Science and Technology in Development Planning, Mexico.

Reuber, G.L., et al 1973. *Private Foreign Investment in Development* (Clarendon)

Sagasti, F. 1978. *Science and Technology for Development,* Main Comparative Report of the Science and Technology Policy Implements Project, I.R.D.C.

Sercovich, F.C. 1974. "Foreign Technology and Control in the Argentinian Industry," (PH.D., Sussex).

Stewart, F. 1977. *Technology and Underdevelopment,* (Macmillan)

_____ . Forthcoming. "Technology Transfer: A Consideration of the Role of Taxation," in *Technology Transfer Control Systems, Issues Prospectives and Implications,* ed. T. Sagafi-nejad, R. Moxon and H.V. Perlmutter.

UNCTAD 1975a. "Major Issues Arising in the Transfer of Technology to Developing Countries," (TD/B/AC.11/10, Rev. 2.)

_____ . 1975 b. "The Role of the Patent System in the Transfer of Technology to Developing Countries" (TD/B/AC. 11/19. Rev. 1.)

UNIDO. 1978. "Recent Developments in the Regulation of Foreign Technology in Selected Developing Countries," (ID/WG.275/8).

Vaitsos, C.V. 1971. "The Process of Commercialization of Technology in the Andean Pact," (O.A.S.).

1974. *Intercountry Income Distribution and Transnational Enterprises,* (O.U.P.)

Vallianatos, E.G. 1976. *Fear in the Countryside, The Control of Agricultural Resources in Poor Countries,* Werman A., (1979) "Planning of (Ballinger) Development, Science and Technology; the Case of the Mexican Agricultural Sector," Symposium on Science and Technology Development Planning, Mexico, 1979.

Wells, L.T. 1978. "Foreign Investment from the Third World: The Experience of Chinese Firms from Hong Kong", *Columbia Journal of World Business.*

_____ . 1979. "Developing Country Investors in Indonesia," *Bulletin* of *Indonesian Economic Studies.*

8

Technology and the Structure of the International System

VICTOR BASIUK*

Introduction:
The Problem of Defining "Structure"

Technological advance, backed by innovations in science, has probably been the single most important factor in societal change in recent centuries. It has been said that, if we take away technological advance, present advanced societies would not differ much from those in the second century B.C. Considering the powerful effect of technology on societal change, it may be appropriate to address the question of the extent of the impact of technology on the international system, with particular reference to its structure.

The subject of this paper is complicated by the fact that there is no agreement among theorists of international relations as to what comprises the structure of the international system. At least three approaches to the question of structure can be differentiated.

Deductively oriented theorists — for example, Kenneth Waltz[1] — attempt to disassociate the structure from the properties and behavior of in-

*I wish to express my thanks to William T. R. Fox and Robert L. Rothstein for comments on this chapter.

dividual units. As Waltz puts it, "to define a structure requires ignoring how units relate with one another (how they interact) and concentrating on how they stand in relation to one another (how they are arranged or positioned)."[2] This approach gives us a high degree of durability and permanency as to what comprises "structure," but it reduces the components of the structure to very few, general, and obvious characteristics of the international system, which do not provide us with a sufficient yardstick to measure change in world politics or adequately explain (let alone predict) the nature and behavior of the units of the international system.

Thus, according to Waltz, the structure of the international system consists of:

1. The ordering principle of the system, i.e., how parts relate to the whole. Here, the answer is reasonably clear and obvious: international systems, unlike domestic systems, are decentralized and anarchic, "formed by the coaction of self-regarding units";
2. The character of the units. Waltz's answer is rather traditional: nation-states are the units of the international system and will long remain so;
3. The distribution of capabilities among the units. If you change the distribution of power among nation-states, you change the structure of the international system.

On the opposite side of the spectrum are some inductively oriented theorists who attempt to find structure through an analysis focusing on changes in the international forum and on the behavior of individual units. This approach is more helpful in explaining behavior in concrete situations and is more policy-relevant. Its liability is that it disperses the search for structure over many variables and often finds structural transformation in events which are important, but short of truly fundamental to be considered as structural changes.[3]

A third group of analysts — who usually focus on new trends and forces in world politics — tend to associate "structure" with power and hierarchical relationship. Apparently influenced by the changing role of power in the international arena, they thus tend to downgrade the importance of structure in the international system or, at least, give it a narrow meaning. Thus, Robert Keohane and Joseph Nye view the structure of a system as "the distribution of capabilities among similar units."[4]

The question of structure is not a trivial one. A major task in the field of international relations is to develop a comprehensive theory that would help elucidate causality in world politics and thus assist the policy maker in steering the nation in a desirable direction at minimal cost. Such a theory inevitably requires the discernment of the structure of the international system, of the process of world politics, and of the nature of the relationship between the two in each period of history.

As Morton Kaplan observed,[5] we are in an early stage of the development of the theory of international relations. In order to advance the state of the art, some of the requirements of the theory must be loosened. Accordingly, in this essay I attempt to strike a balance between a deductive and an inductive approach, by selecting those elements attributable to the behavior of states that assumed a sufficient degree of permanency so as to be viewed as structural elements. The structure is, therefore, defined here as an aggregate of permanent or semi-permanent characteristics of the international system that determines its nature, effects, and, to a degree, governs the daily relations among the actors-participants in the system.

The structural elements considered in this essay are:

1. The principle by which the system is governed (centralization vs. decentralization, supreme authority vs. anarchy).
2. The identity of the units of the system (nation-states vs. new emerging actors in world politics).
3. The number of units in this system.
4. The distribution of power among them.
5. The objectives of the units.
6. The alignments among the units.
7. The rules and customs that govern the relationship among the units.[6]

The impact of technology on each of these elements will be discussed. Since, until recent decades, the greatest impact of technology was on the distribution of power among the units of the international system, that subject will be addressed first and at some length.

Technology and the Distribution of Power

Although, historically, the single most important impact of technology on the structure of the international system was through changes in the distribution of power among its units, the role of technology in this regard should not be exaggerated. Certain political developments had to take place and certain political prerequisites had to be met before technology could exert its influence. This can be illustrated by the three most important changes in the world distribution of power in the past 150 years — the rise of the United States and the Soviet Union to superpower status, and the emergence of Germany in the second half of the nineteenth century and the first half of the twentieth as the leading power in continental Europe. In all the three cases, ideology and/or political will were critical to power, and, in the last analysis, they were more important than technological factors.

In the case of the United States, the ideology of freedom, which attracted masses of immigrants from overseas, was highly important. The combination of the political will and of the economic drive to expand westward created a continent-size nation on which an industrial base of unprecedented magnitude could be built. In the case of the Soviet Union, the role of political will was even more pronounced. The will of Czarist Russia for territorial expansion at the expense of its neighbors, and the will of Communist Russia for developing industrial power resulted in the growth of the Soviet Union to a superpower status. The rise of Germany was a product of the policies of Bismarck and of nationalism, which reached its peak in Nazi Germany.

While political factors, in whatever form, were indispensable for rise in power, the emergence and application of a particular technology were often decisive in the magnitude and direction of power change. In this connection, we can distinguish three forms of technological impact on the distribution of power:

1. As an *independent* variable. Here, a nation-state does not deliberately promote or use technology for power-related objectives, but, aided by a favorable natural resources base, manpower, geographic location, etc., a given technology may contribute to power of the particular nation-state much more than to that of others. In this case, there is a strong deterministic element in the impact of technology on the distribution of power.
2. As a *dependent* variable. In this form, a nation-state deliberately and purposefully uses technology for power, or assists the growth of a particular technology that leads to the augmentation of the nation-state's power.
3. Through an *indirect* impact. In the previous two forms, the impact of technology was direct in the sense that it was in the form of an instrument that was either available or not available to a state in sufficient magnitude for power purposes. However, technological impact may also favor or damage a state indirectly by changing certain attributes of warfare or the international climate. In this instance, technology also produces impact as an independent variable.

Below, I shall provide some specific illustrations how particular technological impacts or innovations contributed to the rise in power of the United States, the Soviet Union (or the Russian Empire, if before 1917), and Germany. Inasmuch as the period under consideration witnessed a decline of Great Britain as the foremost world power of the nineteenth century, appropriate references to the impact of technology on the British position will be made.

Throughout most of the nineteenth cetury — and at least until World War II — steel production was probably the single most important index of power potential of nations. Great Britain had captured an early lead in this area through the production of crucible steel. This was a slow process, requiring careful attention by skilled artisans, and producing relatively small quantities of steel. The invention of the Bessemer process of mass steel production (1856) broke the initial lead of Great Britain in this area. What mattered in mass steel production was the availability of resources, and not the highly specialized skills of artisans.

The United States, riding on the strength of the superiority of its resources (iron ore and coal in close proximity or readily transportable via the Great Lakes) had become the greatest steel producer in the world by 1890. By 1893, Germany had exceeded Great Britain in steel production. The case of the Bessemer process is a vivid example of the impact of technology largely as an independent variable: ironically, the process was a British invention.[7]

Another major example of technology acting largely as an independent variable and contributing to a major change in the world distribution of power was the case of railroads. Here again, through inventions by her subjects,[8] Great Britain was most obliging in undermining her own power position in the long run. Railroads were instrumental in integrating and developing the hinterland of continental nations and endowing them with overland mobility, thus challenging the predominant position of Great Britain in long-distance mobility enjoyed through her seapower. Because of railroads, the power position between Great Britain and Russia was significantly affected in the span of some 50 years, between the mid-nineteenth century and the beginning of the twentieth.

In 1853–56, Great Britain, assisted by allies, won the Crimean War on Russia's own soil largely because Russian railroads were inadequately developed to mobilize the Czar's military power and project it to Crimea. But, at the turn of the century, the British government and public opinion were seriously concerned about the threat to India through Russia's construction of the Trans-Siberian and the Trans-Caspian railroad. The latter, in particular, was aimed in the direction of India. A quotation from a contemporary (1900) magazine provides a picturesque illustration of how the threat of the Russian railroad construction was viewed at the time:

> Like a hand, the palm of which firmly covers Siberia and Transcaspia, the Czar has planted his railway system in Asia. From this strong palm five fingers radiate, and feel their way preparatory to closing in for the next firm grasp. At present an apparently weak and insignificant finger of steel is gradually slipping its point of connection with the palm, in the Caucusus, toward Constantinople; it has made some progress. The next finger almost touches Teheran and is gliding down through Persia to the Gulf. The middle finger slips out from under the palm at Merv, in central Asia; it has touched Herat, and, unless stopped, will soon reach Kandahar and the Arabian Sea, touching British India teasingly in the ribs near Karachi. The index finger, starting from Samarkand, has already reached the border of China. It will penetrate through the center, its tip finally emerging at Peking, that long-coveted plum, the capital of China, on which already rests Russia's thumb, like a powerful lever coming down from Siberia.[9]

The rise of railroads was bound to benefit the continental giants like the

United States, Russia, and Germany. Railroads were critical in the development of America's resources and war potential. Here, their contribution to power was, in part, as an independent and, in part, as a dependent variable, inasmuch as their growth was assisted by the U.S. government.[10] In the case of Germany, railroads not only contributed to the development of her power potential through economic integration of the country, but their skillful use in warfare contributed to direct military victories and corresponding augmentation of Germany's power.

Moltke, the Prussian Chief of the General Staff, was an early student of the potential of railroads in military strategy and effectively used them in the Danish War (1864), the Austro-Prussian War (1866), and the Franco-Prussian War (1870–71).[11] These victories not only consolidated and forged modern Germany, but importantly enhanced her power potential. As a result of the Franco-Prussian War, Germany acquired extensive deposits of Lorraine ore which, because of its high phosphoric content, was thought to be nearly useless at the time. But, the ever-helpful Great Britain obliged, in 1878, with the invention by its subject, Sidney Gilchrist Thomas, of the Thomas process that succeeded in removing phosphorus from the ore during smelting. The Lorraine deposits soon became, by far, the largest source of German iron ore.[12]

In the last 30 years before World War I, technological developments made a highly important *indirect* impact on the world distribution of power by changing the nature of warfare. In turn, this development had far-reaching implications for the structure of the international system.

Towards the end of the nineteenth century, largely unperceived by contemporary observers, there was a shift in importance from the striking power-in-being to power potential in deciding the outcome of wars. Advances in technology were the single most important factor in this development, although there were other important contributing factors.[13]

As a result of the introduction of smokeless powder in 1886 and the subsequent development of rapid-firing artillery,[14] firepower increased tremendously. One effect of this development was to convert what was traditionally regarded as military capital goods into military consumer goods: Not only the ammunition, but the guns were being consumed. In turn, this required industrial power — made possible by technology — to supply continuously implements of war, guns and munitions, in vast quantities.

The increase in fire power led to stalemate at the front. Railroads provided strategic mobility by bringing striking power to the front, but they could not cross the front to provide tactical mobility. The task of crossing the front was left to the individual soldier who — largely unprotected and vulnerable in his slow movements — had to face a vast barage of firepower. The tank and the airplane, which could have provided the requisite tactical mobility, had not been sufficiently developed and broadly introduced by World War I to make an important difference. As a result, the *sitzkrieg* ensued. The eventual victory was decided by the superior power potential.[15]

In terms of power, the greatest beneficiary of this development was the United States for at least two reasons: (1) at a time when power potential began to matter, she had the largest power potential in the world; (2) with wars becoming protracted (it took time for power potential to be exhausted in warfare), the United States gained the essential time for converting her industrial might to war requirements.

If not for the aforementioned technological effects on warfare, Germany would have won World War I before the United States had a chance to step in. This was precisely the situation Great Britain herself faced in the Franco-Prussian War of 1870–71. And a unification of continenal Europe under German hegemony after World War I would have had far-reaching implications for a post-war structure of the international system. As it was, by the end of World War I, the United States had emerged as the most powerful nation on earth.

Although defeated in World War I, Germany was only a partial loser. Insofar as technological developments favored power potential, she was a beneficiary, since Germany still possessed the single largest industrial complex in Europe. Her re-emergence in the 1930s as the leading military power in the world was principally due to her effectiveness in capitalizing on the two technological developments — the aircraft and the tank — that restored tactical mobility and, just as important, gave new dimensions to strategic mobility. Given the conditions of global warfare with its dispersion of power centers, these technological developments were not sufficient to restore the striking power-in-being to its former preeminence. It was the power potential of the allies — mainly that of the United States — that won World War II.

The emergence of nuclear weapons was a revolutionary development in that they made it possible to destroy the very essence of power — power potential itself — in a matter of days, and not years. If and when used, this would mean the closing of the cycle from short wars — based on the striking power-in-being of the pre-World War I years — to protracted wars of power potential, and then back to *blitzkrieg*. From the point of view of world distribution of power and the structure of the international system, nuclear weapons were important in that they enabled the Soviet Union to move rapidly to the forefront of power, in spite of a GNP which was (then) considerably less than a half of that of the United States. Bipolarity thus ensued, a structural characteristic that dominated the international system for about two decades after World War II.

Reluctance to use nuclear weapons, accompanied by the emergence of the doctrine of limited war in a nuclear environment in the late 1950s,[16] kept power potential as a key attribute of power of nation-states. By 1960, war-devastated nations had restored their industrial power. Technology in particular contributed to the re-emergence of Germany and Japan as major economies: with most of their power potential destroyed in World War II, they were not burdened by an obsolescent technological superstructure and, therefore, could capitalize on the most advanced industrial technology to rebuild their economies. Aided by the transfer of technology from advanced nations, new industrial centers began to emerge (e.g., Brazil). In part, as a result of these developments, bipolarity started to decline.

Perhaps even more importantly, technology contributed to the decline of bipolarity in a less direct and more subtle way. In particular, the following took place:

1. the various factors of power in the international arena have changed in their relative importance;
2. the importance of power in general has declined;
3. the nature of international power began to show signs of change.

There has been a decline in the importance of *military* power. Several factors — all related to technological impact — contributed to this. The propensity of nuclear power to result in a stalemate has been generally recognized. The fact that, for a number of reasons, there is also a tendency in the "conventional" military sector toward stalemate has not been widely noted.[17] A stalemate, or a mutual neutralization of military power, and the concomitant reluctance to use it, lead to a decrease in the power's importance.[18]

A deliberate policy on the part of the superpowers — more so on the part of the United States than of the Soviet Union — to use the stalemating attributes of military power in order to achieve international stability was also a contributory factor. The relative success in achieving military stability in itself tended to diminish the importance of military power, even though the relative neglect of conventional military power by the United States in recent years and such events as Afghanistan provoked a ripple of awakening as to its role and importance. Perhaps more salient, with the introduction of a measure of stability in the international arena, political and economic relations and problems have risen in significance and begun increasingly to occupy the attention of nations. Along with them, political and economic factors have grown in importance as instrumentalities of influence and power.

Under the impact of technology, both economic and technological interdependence have grown. The scope for employment of military power, and even for its threat, has narrowed. Military power is unique in that it is highly disruptive to a multiplicity of other relations and exchanges. Therefore, with increased interdependence, the cost of resorting to military force has increased correspondingly. Resort to military force continues, but only where the stakes are high enough to justify the increasingly multiple (and not only military) costs of its use.[19]

The reasons for the decline in the importance of power are also complex. In part, power has declined because of the decline in the importance of military power, which is perhaps the only "true" form of power in the sense that it is geared and organized toward a single end — potential coercion. Unless used for war, the application of other elements of power — political, economic, technological, and psychological — is usually an "imperfect" multivalue process in which values frequently run at cross-currents.[20]

Also contributing to the decline in the importance of power has been pluralization and diffusion of power inside societies and in the world arena. The nation-state (or another international actor) frequently finds it difficult to mobilize an adequate amount of power — of whatever nature and for whatever purpose — to be truly effective in a contemporary multifocal environment of world politics. As will be discussed at greater length later in this study, technological advance contributed to this development through its propensity to differentiate functions and interests and thus pluralize societies.

A number of observers of the international scene noted that there has been a change in the international political process in recent years. In particular, the zero-sum-game — the concept that one's gain was necessarily somebody else's loss and *vice versa* — which prevailed in world politics of earlier years, began to give in to a recognition of the need for a non-zero-sum approach. In a non-zero-sum (or positive sum) game, one party's gain is not necessarily somebody else's loss; both could gain. Technological advance was a primary contributor to the relative rise of the non-zero-sum game in world politics. In nuclear weapons, a gain by one state could result in mutual suicide. To optimize national security, one must cooperate rather than compete. Degradation of the environment because of technological advance also required cooperation across national frontiers, not competition. The huge costs of science and technology suggested advisability of the pooling of resources and cooperation rather than competition, if the multiple problems of contemporary societies were to be solved.

Zero-sum-game is still prevalent in world politics, but the relative gain of the positive-sum-game tended to modify the nature of power in the international forum. Power as "domination" or "control" began to recede, while power as "influence" is becoming increasingly important. Power is also becoming less hierarchical and more issue-relevant and issue-oriented.

In short, largely under the influence of technology, the *process* of world politics and the role and nature of power within it have undergone change. In turn, this affected the *structure* of the international system. In particular, while in terms of military power the world is still bipolar, this is no longer so in terms of the totality of relevant power factors.[21]

Technology and the Alignments Among the Units of the International System

Until relatively recently, technology had very little, if any, *direct* effect on the alignments of the units of the international system. It did, however, play a significant role in influencing the alignments *indirectly,* by changing the distribution of power among the units (nation-states). Once a change in the distribution of power took place or was perceived to be taking place, the shift in the alignments, if any, usually followed in response to the considerations of the balance of power.

As far as the structure of the international system was concerned, it was not so much the change in the distribution of power as the *use* of power and the resultant alignments that were important. The single most important development in terms of change in the distribution of power between the Treaty of Vienna of 1815 and World War I was the rise of the United States as an industrial giant.[22] But the single most important development which changed the structure of the international system in the same period was probably the Franco-Prussian War of 1870–71. Its significance lay not so much in augmenting Germany's power through the spoils of the war, as in triggering a chain of events that transformed the chandelier balance of power of post-Napoleonic years to a polarization of power, culminating in World War I.

The rise of nuclear weapons after World War II stimulated a somewhat different perception on the impact of military technology on unit alignments. A question was raised to the effect that, rather than following the conventional considerations of the balance of power, perhaps it would be safer for certain particularly vulnerable states or regions to assume a posture of neutrality. Although discussed in the press and scholarly publications, this notion did not find its way into formal alliance structures. Even Japan — an island nation, perhaps the most vulnerable to the destructive force of nuclear weapons — has remained a U.S. ally, although some analysts raised the possibility that it may choose neutrality in an actual outbreak of a nuclear war.

The immediate effect of the acquisition of nuclear weapons by both the United States and the Soviet Union was to sharpen bipolarity on a world scale. However, as discussed earlier, by the early 1960s bipolarity began to get diffused in the interplay of nonmilitary power factors. Moreover, as the scale and costs of technology were growing, a new phenomenon began to emerge: the possibility that cooperation in the development and utilization of technology may influence alignments of nation-states.

At the height of détente, there was a distinct expectation in some U.S. private and governmental circles that cooperation in science and technology between the United States and the Soviet Union may, in the long run, significantly change the relationship between the superpowers. In the late 1960s, a proposal by Soviet scientists was broadly discussed in the Soviet and Canadian press which, if implemented, would have established conditions making a war between the United States and the Soviet Union an extremely unlikely event. The proposal, worked out in technical details, involved the feasibility of damming the Bering Strait and, through the pumping of the water into the Pacific, changing the direction of the Gulf Stream and thus ameliorating the climate of Eastern Siberia, Alaska, and Northern Canada.[23] Once these regions had become populated and productive, the destruction of the dam would have been extremely costly to the economies of the United States, the Soviet Union, and Canada.

Such proposals notwithstanding, technology has not been successful so far in changing alignments of nation-states or modifying the relationship between them where political considerations militated against such a development. However, technology has been instrumental in cementing relationships between units of the international system in those cases where political conditions were favorable for such a development or, at least, were not prohibitive. This certainly has been true in the case of Western Europe, where technological cooperation has been a major factor in European integration. A more recent development has been a trans-Atlantic technological cooperation in which the United States usually took initiative. Such cooperation involved pooling technologies and resources of Western European nations and the United States in multinational programs. An example of this activity includes the construction by the European Space Agency (established in 1975) of the Spacelab for NASA's space shuttle at a cost of approximately $875 million (in 1979 dollars).[24] Still another example is the initiation of an effort by the U.S. Department of Defense in 1978 to establish a common coordinated base for military resarch and development of all NATO nations, which would expand the potential technological spectrum, while cutting duplication and cost of R & D.[25]

The role of technology in influencing directly the relationship and alignment of states is likely to increase in the future as the spectrum of potential technologies and the cost of their implementation increasingly exceeds the resources of individual nations. However, one should not overestimate the ability of this trend to overcome serious political and national security obstacles.

A more important recent development involving the impact of technology on the alignment of nation-states is the emergence of the North/South polarization. It is a product of *differential* technological advance and of its politicization. This subject is much too complex to be discussed in any depth here; the reader is, therefore, referred to its treatment elsewhere.[26] Suffice it to say that this development does not necessarily comprise an enduring characteristic of the structure of the international system. In part, its resolution would depend on the ability of the international community — whose problems, in some respects, are beginning to resemble those of domestic societies — to provide measures and avenues facilitating upward mobility for the underprivileged.[27] In part, it would depend on the willingness and ability of the less-developed countries to pursue the most effective means to achieve upward mobility. At this point, the outlook is not particularly promising on both counts.

The International System: A Growing Importance of the Political Process at the Expense of the Structure?

Historically, the contribution of technology to change was probably the greatest in the two areas discussed above: the distribution of power and, secondarily and less directly, alignments among the units of the international system. In the past 25 years or so, some profound changes began to be discernible in the remaining five structural elements enumerated in the introduction to this essay, changes in which technological impact played a significant role. A part of the aggregate of these changes has already been referred to: change in the nature and importance of power. I shall briefly discuss others below.

The impact of technological advance raised the question of the *identity of the units of the international system.* In earlier decades, it was generally accepted that the international system consisted of nation-states, sovereign and equal under international law, if not equal in power. Technological advance, however, produces the following effects on societies as applicable to the role of the nation-state in the international system:

1. It introduces a high degree of complexity by promoting specialization and the need for sustained adherence to a particular role or function on the part of individuals and groups.
2. It promotes the growth of particularistic interests usually associated with proliferating roles and functions; or, to put it differently, technological advance *pluralizes* societies.
3. A number of these interests cut across national frontiers and cooperate among themselves, thus at least partially escaping or by-passing control by the nation-state.
4. Even where major interests are largely confined within national boundaries (and increasingly fewer and fewer of such interests are found in this category), the top leaders of a nation-state often find it difficult to override them because of the interests' high technical complexity. The top leaders simply cannot be confident that they know enough to override the arguments presented to them by the interests involved. The decision-

making thus tends to politicize, with the outcome often determined by the political process. This phenomenon gives certain major interests the extent of influence and power — at times, at the expense of the nation-state — which did not exist before.

5. Because of the magnitude and scope of problems created or affected by technological advance (e.g., nuclear weapons, ecology), nation-states can no longer cope with them separately. Technology also offers a number of advantages through cooperative efforts. Hence the rise of the non-zero-sum political process referred to earlier, as distinguished from the earlier predominance of the zero-sum-game in world politics.

The upshot of this development has been a *diffusion* of power within nations and in the international arena.[28] Contrary to David Mitrany,[29] the principal early exponent of functionalism, the nation-state has not lost most of its authority to powerful "nonpolitical" international institutions in the economic, social, and humanitarian fields which, by fulfilling basic needs of society, would gradually take over many of the functions currently carried out by nation-states. The nation-state has still remained — and is likely to remain for quite some time[30] — the principal actor in world politics, but other actors — regional organizations, the multinational corporation, certain global functional organizations — have proliferated and, in some respects, have curtailed the authority of the nation-state.[31]

In theory, recent advances in technology could radicallly change *the number of units in the international system:* An outbreak of a global nuclear war could literally wipe some nation-states from the map, and if an autocratic state were victorious, this could, conceivably, lead to a near-global empire. Such an outcome, however, is not very likely. But if we extend the concept of units of the international system beyond the nation-state itself to secondary actors in world politics, then we could say that technological impact certainly contributed to the proliferation of such units. This has been discussed in the preceding paragraphs and need not be repeated here.

As interests and interest groups proliferated and began to cut across national frontiers, the separation between foreign and domestic policy came to be increasingly blurred. This, in turn, changed *the nature of the states' objectives.* In the past, foreign policy was the preserve of the diplomat and of the soldier and its essence was the integration of force and policy or the use of coercion on behalf of external goals. These aspects still remain, but they have been fused with the projection of many domestic considerations on the international arena. Nearly every ministry and agency has a foreign policy of its own and attempts to project it abroad. Thus, in addition to the traditional aspects, world politics increasingly becomes an arena for conflict and cooperation between economic and social policies.[32] There is an important difference in degree in this respect between Western democratic states on the one hand and such autocratic states as the Soviet Union on the other. The latter adheres more strongly to the traditional objectives of policy — projection of power and the use of coercion on behalf of external goals. But, even the Soviet Union shows distinct signs of internal pluralization projected into foreign policy.[33]

Technological impact affected *the principle by which the international system is governed,* but it did not radically change it. Although the existence of nuclear weapons created the theoretical potential of a world empire under the hegemony of an autocratic state, in actuality the lack of supreme authority and international anarchy have not only been preserved, but increased. The expectations of the earlier theorists of international relations and law who envisioned a logical progression from the principle of the balance of power to the League of Nations, the United Nations, and, eventually, to a world government proved to be unfounded. The mainstream of international power bypasses the United Nations. If anything, the rise of nuclear weapons reinforced this development: Nuclear powers are hardly enthusiastic about surrendering even the slightest degree of authority over their awesome weapons to an international body that they tend to view as little more than a debating society, not always responsible at that.

Not having established a central authority through conquest or consent, and being subjected to the pluralizing impact of technological advance, world society proceeds in the direction of greater diffusion of power. Although the international system consists of nation-states of unequal power, relationships in the global arena tend to be less hierarchical and more symmetrical, in which a superpower can be stymied by the capture of its embassy personnel by an unstable and weak state. Proliferation of interests inside societies and internationally, and the frequently highly politicized interaction among them, further promotes symmetrical (as distinguished from hierarchical) relationships. Power tends to be not only diffuse, but at times amorphous. While structural elements of the system still exist and can be discerned, they tend to be subdued and, at times, submerged in what appears to be the growing importance of the political process (diffuse and somewhat amorphous, like power) as the prevalent characteristic of the international system.[34]

But how does a system which lacks a central authority, whose power and political process are diffuse and even amorphous, and which is increasingly characterized by overpoliticized symmetrical relationships among multiple actors and interests, manage to govern itself? Indeed, this may be a miracle in its own right, and it brings us to the consideration of our last structural element of the system: *The rules and customs that govern the relationship among the units.*

The concept that has become fashionable among theorists of international relations in recent years is that of "international regimes," defined by Donald Puchala and Raymond Hopkins as "a set of rules, norms or institutional procedures that govern a social system."[35] As Oran Young points out[36] they are basically social structures, which include a substantive component (a collection of rights and rules) and a procedural component (recognized arrangements for resolving situations requiring social or collective choices) and are focused on a particular area of activity, like deep seabed mining, control of armaments, whaling, global international air transport, or international monetary matters.

International regimes are not exactly a new phenomenon — they can be traced back into history. What is new is their proliferation in recent decades and the multiple actors involved in many of them. Formally, the members of international regimes are always nation-states, but they are usually represented by various specialized ministries or agencies — commerce, communications, fishing, etc. Moreover, the parties carrying out the actions governed by international regimes are often private parties (like banks or airlines) and they influence the outcome of decisions of international regimes. In addition, specialized nongovernmental organizations (NGOs) take keen interest in the activities of international regimes and influence them. In short, international regimes reflect, in microcosm, many of the changes in world politics of recent decades, discussed in the preceding pages of this essay.[37]

Some international regimes have very little to do with technology, but many are a direct product of technological advance. The evolving regime of deep seabed mining emerged in direct response to the technological capability for exploiting deep ocean resources. International Civil Aviation Organization (ICAO), which encompasses the international regime on civil aviation, arose in response to a technological innovation that enabled man to fly. International Telecommunications Union (ITU) was also established in response to technological advance, as was the regime on nuclear nonproliferation. In the case of other regimes — like fisheries — the impact of technology was less pronounced, but important nonetheless. Probably, it would not be an exaggeration to say that the impact of technology was a highly important causal factor in the establishment of the majority of international regimes.

Conclusion: *Quo Vadimus?*

In the past, the international system consisted — and was perceived to consist of — nation-states, which interacted somewhat like billiard balls. They were hitting each other in wars and in clashes of interests. They were changing their positions on the table in response to the considerations of the balance of power and of their own ambitions in the international arena. Although separation between foreign and domestic affairs was never truly complete, it was, nonetheless, prevalent. Because of that separation, the impact of technology was much greater on domestic societies than on the structure of the international system. As far as the latter was concerned, technology affected nation-states by being a highly important factor in changing their power, and, indirectly, their position on the billiard table.

To be sure, technology has been integrating the world and creating interdependence within communities and among them for a long time. It was thus affecting the international system, but the impact was slowed down and constrained by the characteristics of the system's structure: "the coaction of self-regarding units," as Kenneth Waltz felicitously put it. The diplomat and the soldier persisted in viewing foreign policy as their preserve, often in spite of the pent-up forces stimulated by technological advance, which demanded, dictated, or pressed for a different approach.

The changes induced by technology began to affect the international system at an unprecedented rate in the past two decades. In particular, the pluralizing impact of technology which profoundly affected individual societies has been extended to the international arena. The hard exterior of the billiard ball had been showing cracks before, but now the cracks widened, with various interests increasingly emerging through them and proceeding to interact on the international forum as if they belonged there. They have thus joined other actors — international organizations — which the nation-states were creating in recognition of their inadequacy in governing the increasingly complex, interdependent world. The distinction between "domestic" and "foreign" affairs has not only become blurred, but, in many respects, obliterated.

What we appear to have witnessed in the last 20 years is the *transition from the international system to a world system*. The impact of technological advance was the single most important factor that has brought about this transition.

Is the change from the international system to an inchoate world system a step forward, an improvement? At best, one could say it is a reflection of reality brought about by technological advance. But it is not clear where this reality is taking us. For a student of government and political science who would like to see a rational, reasonably stable and reasonably well-governed world, the picture is hardly reassuring. The old hopes for the rise of a hierarchical structure on the global arena proved to be illusory. The U.N. system, although comprehensive and global in scope, has evolved into a number of "international regimes," coping with a multiplicity of issue areas. Indeed, international regimes, within the U.N. system and without, have emerged as the mainstay of world order. Considering that their effectiveness is highly uneven, the relationship between them is loose and sometimes non-existent, and their activities are often highly politicized, the regimes hardly represent a neat, orderly structure whose evolution carries a promise of a better world in the future.

In a pluralistic world increasingly pluralized and complicated by the impact of technology, problems abound, proliferate, and tend to politicize. Trying to cope with various problems as they come, in a semi-compartmented way and somewhat on an *ad hoc* basis, might well prove to be an impossible task, which would only demonstrate that the inchoate world system is incapable of providing the minimum of stability and order essential for the security, growth, and viability of its various components.

What does this suggest for both the scholar and the practitioner in foreign affairs? Perhaps a promising approach lies not in focusing on problems and problem-solving — a process that may engulf us all — but on developing concepts and leverages of power that would give the world system a better sense of direction for its evolution and, eventually, a structure capable of sustaining and enhancing its viability. For the theorist in international relations, this might require a discovery of a healthy combination of the deductive and the normative approaches, yet solidly supported by empirical research.

This is a large order and it cannot be filled readily or very soon. The analysis of this paper also suggests concerns and questions closer to home. In a world system whose organization cannot safeguard America's security and ensure its viability, important questions arise with regard to the well-being of the United States. Considering the magnitude and complexity of technological impact on the United States and the world, are our policy makers sufficiently appreciative of its many ramifications and possibilities to guide them in their decisions and the formulation of requisite policies? Do we have a comprehensive and sufficiently foresighted policy with regard to technology transfer, so as to ensure America's viability for this and future generations? Or are we, like Great Britain in the past, helping our present and potential future rivals? Unless we satisfactorily answer these and similar questions, we'll hardly be in a position to mold the structure of the inchoate world system.

Notes

1. Kenneth N. Waltz, *Theory of International Politics* (Reading, Mass.: Addison-Wesley Publishing Co., 1979), especially pp. 79-101.

2. *Ibid.* p. 80

3. For example, Charles Doran views the five-fold increase in oil prices in 1973 by OPEC as a case in system transformation, involving structural change. Although Doran presents a reasoned analysis of the importance of this development in a historical perspective, it is not a convincing case of system transformation. See Charles F. Doran, "Modes, Mechanisms, and Turning Points: Perspectives on Systems Transformation," a paper delivered at the 1979 Annual Meeting of the American Political Science Association, Washington, D.C. August 30 — September 2, 1979, pp 27-28.

4. Robert O. Keohane and Joseph S. Nye, *Power and Interdependence; World Politics in Transition* (Boston: Little, Brown and Co., 1977), p. 20. A similar view appears to be shared by

Stanley Hoffmann. His *Primacy or World Order; American Foreign Policy Since the Cold War* (New York: McGraw-Hill Book Co., 1978) does not address itself at any length to the question of structure of the international system. At one point, noting a change in the number and nature of actors, he observes: "This is one of the reasons why a purely 'structural' analysis of the international system that focuses only on the number of great powers and on 'the distribution of capabilities among units' tells us little." (p. 111).

5. Morton A. Kaplan, *Towards Professionalism in International Theory* (New York: The Free Press, 1979), p. 130.

6. These elements are similar to, but not identical with, those compiled by Dina A. Zinnes in "The Mystery of the Missing Idea: A Non Paper on System Transformations," Reports and Research Studies, Center for International Policy Studies, Indiana University, Bloomington, Indiana, August 1979, pp. 8–17.

7. For details, see Victor Basiuk, "The Differential Impact of Technological Change on the Great Powers, 1870-1914; The Case of Steel" (Ph.D. Dissertation, Columbia University, 1956), pp. 11–19. (Hereafter cited as Basiuk, "The Differential Impact of Technological Change").

8. The steam locomotive capable of heavy haulage was invented by Richard Trevithick in 1804. The first steam-powered railroad was built in England in 1825.

9. Alexander Ford, "The Warfare of Railways in Asia," *Century Magazine,* Vol. LIX (March, 1900), p. 794. See also Basiuk, "The Differential Impact of Technological Change," pp. 165–181, on Russia's use of railroads for power.

10. For details, see *ibid,* pp. 81–139.

11. Hajo Holborn, "Moltke and Schlieffen: The Prussian-German School," in Edward Mead Earle (ed.), *Makers of Modern Strategy* (Princeton: Princeton University Press, 1943), pp. 177–178.

12. The value of the Thomas process to Germany was considerably greater than its contribution to steel production. The phosphates derived as a byproduct were important for German agricultural development as a fertilizer. See Basiuk, "The Differential Impact of Technological Change," pp. 36–41.

13. These other factors included a general introduction of conscription after the Franco-Prussian War of 1870–71 among Europen powers and the resultant rise of mass armies. Mass armies shortened the link between power potential and military power. The rise of nationalism was another factor.

14. There was no incentive to develop rapid-firing artillery in the era of black powder for the simple reason that, after several shots, there was so much smoke in the air that the crew could not see where to shoot.

15. For a more extensive treatment of this subject, see Basiuk, "The Differential Impact of Technological Change," pp. 218–273; 363–366.

16. See Henry A. Kissinger, *Nuclear Weapons and Foreign Policy* (New York: Council on Foreign Relations, 1957) and Robert E. Osgood, *Limited War; The Challenge to American Strategy* (Chicago: University of Chicago Press, 1957).

17. See Victor Basiuk, *Technology, World Politics, and American Policy* (New York: Columbia University Press, 1977), pp. 15–16. (Hereafter referred to as Basiuk, *Technology, World Politics).*

18. However, when military power is neglected, becomes seriously out of balance, and fails in its stalemating role, it rises in importance again. This is precisely what has happened in recent years in connection with the inadequacy of U.S. military power for projection to the Persian Gulf.

19. In the case of Afghanistan — and perhaps beyond Soviet expectations — the Soviet Union discovered the multiple costs of the use of military force. The multiple costs to the Soviet Union of the potential military intervention in Poland were continuously being pointed out by the United States and its allies in November and December of 1980 as a means of deterrence.

20. The political process is basically a bargaining process in which considerations of long-range accommodation may be more important than an immediate power gain. Economic factors are primarily concerned with production of goods and satisfaction of human wants; when they are employed for power purposes, their role, in a sense, is artificially distorted. The same is true of technological and psychological factors of power. See Basiuk, *Technology, World Politics,* pp. 151-152.

21. The above discussion of power is based largely on *ibid,* pp. 150-153, 187-189; *cf.* also Hoffmann, *op.cit.,* pp. 114-119.

22. One might argue that the rise of the United States produced little, if any, impact on unit alignments before World War I because the United States was not involved in the then balance of power process. This, at best, could be a partial and not a completely satisfactory explanation. The United States became involved in what was developing into a global balance of power process through its participation in Far Eastern affairs around the turn of the century. And, as the polarization among the Great Powers sharpened in pre-World War I years, U.S. direct involvement in the European balance of power could have been anticipated. The key element affecting the structure of the international system then appears to be the way in which power is actually used, and not its mere existence. This suggests that, in this respect, political rather than technological factors are by far the more important.

23. See P. M. Borisov, "Can We Control the Arctic Climate?" *Bulletin of the Atomic Scientists,* Vol. 25 (March 1969), pp. 43-48 and Ye. Gushchenkov, "Can the Pacific Warm the Arctic?" *Ibid.* (May, 1969), pp. 69ff.

24. Source: NASA.

25. See The Hon. William J. Perry, Under Secretary of Defense for Research and Engineering, "The Department of Defense Written Statement of NATO-Improved Armaments Cooperation to the Research and Development Subcommittee of the Committee on Armed Services of the U.S. Senate" (96th Congress, First Session, April 4, 1979), pp. 6-19. The objective of this effort is to yield $16 to $17 billion worth of combined results in R & D ($12 billion spent by the United States and $4 to $5 billion spent by U.S. NATO allies).

26. See Robert L. Rothstein, *The Weak in the World of the Strong; The Developing Countries in the International System* (New York: Columbia U. Press, 1977).

27. I am indebted to William T. R. Fox for this observation.

28. See Basiuk, *Technology, World Politics,* pp. 190-192.

29. *A Working Peace System* (Chicago: Quadrangle Books, 1966). This is a reprint of Mitrany's publications of the previous two decades.

30. This is primarily so because there is no sufficiently effective substitute in sight for the single most important function of the nation-state: to protect human values and to help satisfy value goals. In its two other functions — (1) to provide organization for change, in response to changing values or material, political or other conditions, and (2) to provide organization for stability — the nation-state is becoming less effective and is assisted by international organization.

31. Stanley Hoffmann *(op.cit.,* pp. 111-112) suggests that "nonstate actors, even though they may not be 'sovereign' . . . can nevertheless be players. As such, they can be considered part of the system's 'structure' and not merely of the 'processes that go on within it.' . . . They are endowed with power over resources essential to the state; they can bargain directly with the legitimate wielders of political power (or try to undermine them); and even when they are a mere emanation of a state, or a creation of a state, they have a *de facto* autonomy within certain ranges."

32. *Cf.* Hoffmann, *op.cit.,* p. 113.

33. See Victor Basiuk, "Implications of Differential Transfer of Technology to the USSR and Resultant Options for U.S. Technology Transfer Policy"; Report to the Deputy Under Secretary of Defense (International Programs and Technology). McLean, Virginia, March

1981, pp. 7-9. For example, the Soviet Ministry of Foreign Trade, in obvious self-interest, was strongly in favor of technology transfer from the West during the debate on this subject in the late 1960s and the early 1970s and continued to extoll the virtues of Western technology and the principle of comparative advantage in its publications in later years, a position which was at variance with some other Soviet agencies.

34. Facing problems in finding a coherent over-all structure of the international system, Keohane and Nye *(op.cit.,* pp. 50-51) suggest "issue structuralism" as an approach to the understanding of contemporary world politics. In issue structuralism, power of states is assessed with respect to a given issue area. This approach is useful as a tool of analysis in *ad hoc* situations or issue areas, but it does not quite provide a substitute for the need of a more comprehensive theory discerning an overall structure, the nature of the political process, and the relationship between the two.

35. Donald J. Puchala and Raymond F. Hopkins, "Regimes and Political Theory: Lessons from Inductive Analysis"; Paper delivered at the 1980 Annual Meeting of the American Political Science Association, Washington, D.C., August 28-31, p. 1.

36. Oran R. Young, "International Regimes: Problems of Concept Formation," *World Politics,* Vol. XXXII (April 1980), pp. 332-337.

37. On this subject, *cf.* also Ernst B. Haas, "Why Collaborate? Issue Linkage and International Regimes," *World Politics,* Vol. XXXII (April, 1980), pp. 357-405.

9

Letting the Genie
Out of the Bottle?

The Micro-electronic Revolution
and North-South Relations
in the 1980s

WARD MOREHOUSE

Parts of this paper are drawn from some background notes which I prepared for the Lund Workshop on Technological Change in Industrialized Countries and its Consequences for Developing and Industrialized Countries, "Closing the Circle: The Question of Technological Choice and Employment in Industrialized Countries," and from a summary of the Workshop prepared for the *Lund Letter on Science, Technology, and Basic Human Needs,* No. 13 (June 1979). My indebtedness to the other participants in the Workshop, which was organized by the Research Policy Institute at the University of Lund on 28–30 May 1979, and to other colleagues at the Institute is even more substantial than the citations to the paper suggest. (The presentations to and notes for the Lund Workshop were for the most part based on work in progress. They have been cited here in order to give proper credit for a number of points in this paper, but it should be emphasized that, as is the nature of work in progress, they are subject to changes and refinement by their respective authors.)

A number of persons were kind enough to read drafts of this paper and make useful comments, including John Bessant, Ray Curnow, Dieter Ernst, Staffan Jacobsson, Raphael Kaplinsky, Vladimir Pertot, Maurice Strong, Joseph Szyliowicz, and Louis Turner. I am grateful to all of them for their ideas and insights.

Needless to say, none of the foregoing is responsible for any of the unlikely, perhaps even untenable, propositions advanced here, with which some of them almost certainly disagree.

The Basic Thesis

The genie of revolutionary technological change is at least head and shoulders already out of the bottle as we enter the final two decades of the twentieth century. Whether the genie gets out the rest of the way will depend on a variety of economic and political factors, not technical. Technologically, the genie is already sufficiently well formed to assure profound economic, social, and political impact and conflict if it gets all the way out of the bottle.

History tells us that bottling up technological genies is extraordinarily difficult, some would say impossible, in the long run. The genie in this case is the revolution in micro-electronic technology already beginning to hit the industrialized economies. When it strikes those economies with full force, it will alter substantially the global political economy, affecting developing countries as well. The critical questions are the rate at which the micro-electronic genie emerges from the bottle and the political and economic responses it generates within and among nations in the modern world. It should be emphasized at the outset in arguing this thesis that, because the micro-electronic genie is not yet all the way out of the bottle, we know all too little about its actual social, economic, and political consequences. Research into this vitally important issue is still at an embryonic stage. Therefore, many of the assertions I have made here should be read as hypotheses for which there is certainly some evidence available, but which equally certainly require additional validation through further collection and analysis of empirical data. We need collection and analysis of data both at the micro-level, in order to build up the number of cases on which generalizations can be based, and at the macro-level in order to identify broad trend lines. But it is also worth emphasizing that the basic trends on which this thesis rests are already apparent in both industrialized and developing countries. If the micro-electronic genie gets out of the bottle quickly — say, by the middle of the next decade, or even the end of that decade — these trends will become sharply accentuated. Such an observation about the social impact of technology is always subject to the qualification that political and economic factors, having created the conditions that led to the creation of the genie in the first place, can sharply alter its consequences for society as well.

More speculative is a second revolution in materials technology, apparently not much farther in the future. Some would, indeed, argue that, because of the impetus posed by growing shortages and rising prices of non-renewable natural resources, especially for energy, the revolution in materials technology may be much closer than is now realized.[1] The solar transition, in this view, has already begun. Enzyme engineering, as another example, appears to be on the threshold of a major breakthrough, with developments in other areas of biotechnology not far behind. But the revolution in micro-electronics is closest at hand, and should be at the center of international debate on the future of the global political economy. What is striking is its absence from that debate, at least thus far.[2] One of the instrumental purposes of exercises like this paper is, therefore, to stimulate interest and discussion among scholars and policy makers, even at the price of overstating the case. Sooner or later, preferably sooner, this discussion must also involve those whose lives are going to be most affected by these revolutionary technological changes.

The Revolution in Micro-Electronic Technology

We are on the threshold of radical changes in the way in which goods and services are produced and distributed and information is managed and disseminated through micro-electronic technology, as well as, possibly, in the production of radically new products. Whether these changes constitute a "technological revolution" is naturally the subject of some difference of opinion, but there is a substantial number of commentators and analysts who contend that this is the case.[3]

Regardless of the label used, changes in micro-electronics have been so rapid in recent years, and are resulting in such significant cost reductions and performance improvements, that they are bound to have great impact on society to the extent they become widely diffused and adopted. Computing costs, by way of concrete illustration, have dropped from £100 per logic gate in 1960 to about 1 pence by the mid 1970s, and are expected to reach .01 pence by the early 1980s. Equally dramatic increases in computing power, measured in components per chip, have occurred during the same period, rising from 1 in 1960 to 64,000 by 1975 and 1 million by 1980. The figure is expected to reach as much as 16 million by the mid-1980s.[4] Figure 9.1 gives a graphic picture of these intersecting trend lines.

Micro-electronic technology offers a number of advantages in addition to cost reduction, among them speed, flexibility, reliability, maintainability, and the capacity for self-diagnosis, and for being linked to information hierarchies.

The range of applications of micro-electronics is very broad, covering a wide range of goods and functions, many of the latter now being performed by human beings, in a number of manufacturing and service sectors of industrialized economies. Table 9.1 gives concrete examples. Because the advent of really broad diffusion and application of micro-electronics is so close at hand and has been, as a result, so little studied, extent and consequences of using micro-electronic technology in the various products and activities

FIGURE 9.1   Trends in Computing Cost and Power

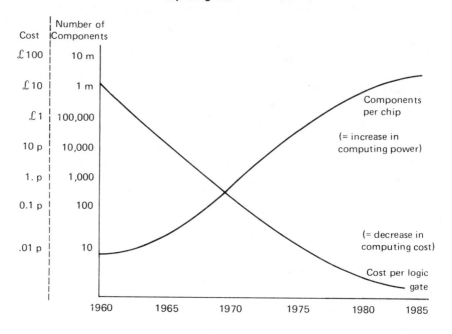

Source: Jon Bessant, Technology Policy Unit, University of Aston, Birmingham, 1979.

noted in Table 9.1 are necessarily hypothetical. Nonetheless, it seems likely that micro-electronics will:

— enhance greatly product differentiation at relatively low incremental cost;
— by reducing economies of scale, make possible decentralization of production, moving production closer to the end user;
— make possible, some would say demand, systems engineering rather than component engineering;
— cut waste and improve efficiency in use of raw materials and energy, in some instance quite substantially;
— reduce variability of final product quality;
— increase, in some cases markedly, the productivity of either or both capital and labor.[5]

Some of these consequences apply particularly to manufacturing, where, in some sectors, the application of micro-electronics may be decisive in maintaining or improving international competitiveness and potentially dramatic

TABLE 9.1   Typical Applications of Microprocessors

Industrial:	Process-control equipment Machine control for machine tools, etc. Material-Handling Systems Robotic systems
Instrumentation:	Analyzers and testing equipment Chemical and gas-measuring systems Optical character recognition systems Oscilloscopes
Commercial and Administrative:	Data-processing equipment Word-processing equipment Banking terminals Cash dispensers Point-of-sale terminals Keyboard terminals for data entry Automatic time clocks and payroll systems Desk-top computers and calculators
Telecommunications:	Telephone exchange systems Telex switching equipment Viewdata systems Satellite communication systems
Health and Educational:	Medical diagnostic and analytical equipment Computer-aided instruction systems Library-control systems
Transportation:	Engine ignition and exhaust control systems Dashboard display systems Braking control systems Fuel pump controls Traffic-control systems Diagnostic systems Aviation systems Aviation applications
Domestic and Personal:	Washing machines, ovens, sewing machines Television, other video and audio systems Cameras, watches Pocket calculators, personal computers

Source: Keith Dickson and John Marsh, *The Micro-Electronics Revolution: A Brief Assessment of the Industrial Impact with a Selected Bibliography,* Birmingham: University of Aston, Technology Policy Unit, *Occasional Paper,* December 1978.

in revitalizing a whole industry. The machine tool industry, in the doldrums in most industrialized countries in the 1970s, now appears to be coming to life with major advances in being (CNC or computer numerical control of machine tools) or in prospect (industrial robots, DNC or direct numerical control of a set of CNC machine tools working in sequence), leading to fully automated factories by the mid to late 1980s, especially in countries with high labor costs.[6]

Notwithstanding the impact on manufacturing, far greater impact is expected in the information sector (including those parts of industrial enterprises concerned with information processing, such as marketing, finance, and management) and on the principal occupational groups in this sector — secretaries, clerks, typists, and managers. Most of these occupations have a substantial proportion of women, giving the impact of micro-electronics a markedly sexist bias. Since the information occupations now constitute the largest single category of the work force in the industrialized countries, at least the more advanced ones (see the figure on trends in composition of the U.S. work force since 1860 by way of illustration), even moderate improvements in productivity could have quite startling results over a short period of time in employment.

Thus, it has been argued that unemployment levels of 10 to 20 percent by the mid-1980s in some of the industrially advanced countries, are by no means out of the question with the application of micro-electronic technology in the information sector by the middle of the next decade, *unless* they are offset by increases in other areas of the work force or they are inhibited by major public policy interventions, worker resistance, or cyclical economic conditions.[7] The point is clearly debatable and I return to it below.

The conclusion of one analyst of the economic and social impact of micro-electronics is that "the overall consequences are comparable with the industrial revolution."[8] With any complex set of phenomena still unfolding, there naturally are others who hold very different views. Optimists take the position that, while the diffusion of micro-electronics may be substantial, the consequences in terms of economic, social, and political adjustments will all sort themselves out without undue trauma. Others take the position that micro-computers are being "oversold" and that in various industrial applications they will be far from cost competitive. They also point out that there are considerable problems yet to be sorted out with the supporting technologies, particularly in software development.[9]

Indeed, it is generally recognized that software will constitute one of the major bottlenecks to widespread diffusion of micro-electronic technology. Although changes in the hardware technology are very dynamic, even volatile — reflected in the rapid cost reductions and increases in computer power noted above — considerable obstacles remain to be overcome before micro-electronics can reach its full potential. Important among these is moving to more accessible computer languages. It appears that we are some distance from that stage, but with such enormous stakes in getting there, it is likely that significant progress will be made in the years immediately ahead.

FIGURE 9.2   Four Sector Aggregation of the U.S. Workforce
By Percent, 1860–1980 (Using modern estimates of information workers)

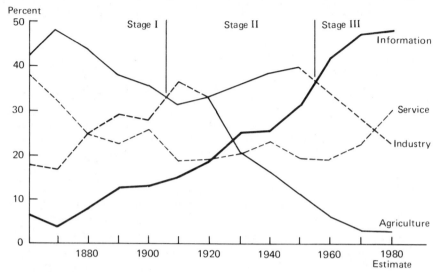

Source: M. U. Porat, *The Information Economy,* Washington: U.S. Department of Commerce, Office of Telecommunications Special Publication (Vol. 77-12-1), May 1977. (Figure is based on U.S. Census Bureau and U.S. Bureau of Labor Statistics data.)

The critical issues in the micro-electronic revolution are not, however, technical but economic, social, and political. It is certainly these factors, rather than technological ones, that will mainly determine the future of micro-electronic technology. Among the most important in terms of its impact on individuals in any social order is the opportunity for meaningful and productive work. To this central issue we shall now turn our attention, examining first the character of conventional wisdom on employment and choice of technique, and then the likely shape of employment in the industrialized countries and in developing countries in the 1980s.

Conventional Wisdom on Employment and Choice of Technique

Conventional wisdom has it that massive unemployment through wrong choice of technique is a problem peculiar to developing countries, although Staffan Jacobsson, in a paper for a recent workshop at the Research Policy Institute in Lund, goes far to challenge that wisdom with respect to the industrial sector of developing country economies.[10] He goes on to show, as have others (e.g., various participants in the November 1978 Bonn Workshop on Technological Dependence such as Seifeddine Bennaceur and Francois Geze of BIPE/Paris in the electronics industry, Pierre Judet of the Institutede Recherche Economique et de Planification (IREP)/Grenoble in iron and steel, and Jacoues Perrin, also of IREP, in engineering), how recent and prospective technological changes are likely to aggravate the problem of employment generation through utilization in developing countries of increasingly capital-intensive, labor-saving technologies generated in industrialized countries to respond to a different mix of technical, factor proportion, market, and other conditions in the latter countries.[11]

The prevailing assumption has been that this issue did not really pose a significant problem for the industrialized economies, aside from temporary adjustments to accommodate new technologies. These economies have generally been regarded as strong enough and resilient enough that the question of "technological unemployment" would take care of itself. After all, in recent decades, certainly since the Second World War, major technological changes appear on balance to have generated at least as many jobs (often in the manufacture, and especially in distribution and servicing of new products based on or derived from new technologies) as they have destroyed. To the extent that new technologies have increased labor productivity in agriculture and manufacturing and thereby displaced labor, this displaced labor has been absorbed in the rapidly growing service sector, which has come to be regarded as one of the primary characteristics of post-industrial societies.

But, is this assumption still valid for industrialized economies? Are we entering a period in which there will be, to borrow a concept from Thomas Kuhn's work on the structure of scientific revolutions, a "paradigm-shift" in the way in which the political economies of the industrialized countries respond to or are affected by major technological changes that now appear to be in prospect or are already in process? The evidence is mixed and not conclusive. But there are some significant trends and compelling hypotheses at the macro-economic level, buttressed by a small but growing body of work at the sectoral level (e.g., the work by Bernard Real of IREP/Grenoble on the machine tool industry and by J.R. Bessant of the Technology Policy Unit at the University of Aston/Birmingham on micro-electronics reported at the recent Lund Workshop and cited above) to suggest that past patterns of job displacement and generation will not necessarily repeat themselves in the foreseeable future (say, the next quarter century), and that industrialized countries may also be confronted with widespread *and* persisting technological unemployment unless there are substantial political interventions in the economy. These political or policy interventions may also have significant impact on choice of technique in industrialized economies. At a minimum, it would appear that technology is a variable to be given more weight in political economic calculations about the future performance of industrialized societies.

In any event, the question of how industrialized societies are affected by and may try to affect the direction of technological change becomes important if there is any substance to the thesis that the future prospects of developing countries are going to be decisively influenced by the nature of the technological change in industrialized societies. I argue below that the question of choice of technique and its impact on employment in industrialized countries is likely to have a substantial impact on the character of North-South relations in the 1980s and beyond.

A "Paradigm-Shift" in the Impact of Technological Change on Employment in Industrialized Societies?

The evidence that we may be, or soon will be experiencing a "paradigm-shift" or fundamental change in the way in which employment in industrialized countries is affected by technological innovation is, as we have suggested, mixed. Christopher Freeman, in summarizing the nature of that evidence growing out of his recent work on technical change and employment, reports two studies on micro-electronics in Germany and Britain which come apparently to opposite conclusions.[12] The German study emphasizes possible increases in employment arising from new products based on micro-electronics (e.g., the toy industry), while the British study underscores that labor-saving techniques would, in the future, increasingly affect the service sector of the economy (historically, the sector which absorbed workers displaced by technical change from increased labor productivity in manufacturing and agriculture) even more than in manufacturing.

Freeman points out that one reason why economists may have underestimated the influence of technological change on unemployment is the tendency to ignore technical change as a significant factor in cyclical fluctuations in the economy. He argues that there are important interactions between cyclical fluctuations and the direction of technological change. Without repeating here the analysis which Freeman gives of the work of a number of economists, he concludes that most economists have not given sufficient weight to the impact of technical change on employment.[13]

An important element in Freeman's argument for giving increased weight to technological innovation in affecting employment in industrialized societies in the future is his concern with "long waves" in world economic development. While this proposition is by no means universally accepted, Freeman concludes that "there is a fairly good case for accepting the reality of long waves in world economic development, with upswings and downswings lasting approximately a quarter century."[14] He is, further, of the view that Schumpeter was essentiallly right in his belief that "the timing of major technical innovations had something to do with the long waves, and in identifying such major changes as steam power, railways, electricity and the internal combustion engine as having repercussions throughout the economic and social system."[15]

Yet another key element in Freeman's analysis is the long time lag involved in the acceptance and diffusion of major new technologies. Before a new technology can really gather momentum and become a substantial factor in the performance of the economy, a great deal of preliminary scientific and technical work must have demonstrated feasibility. As the new technology does gather momentum and its feasibility is more and more widely accepted, relative prices of the goods and services delivered by the new technologies begin to fall and their cost advantage increases. The same conditions apply to secondary industries and services affected by the availability of the new technology. This, according to Freeman, is the significance for microelectronics today of developments for the computer industry during the 1980s.

After a quarter of a century, more or less, the new technologies and branches of industry based on them are well established, and their capacity to generate additional jobs declines sharply. But their employment-displacing effect is still being felt, through a ripple phenomenon elsewhere in the economy. In sum, Freeman argues that the development and diffusion of major new technologies is a long-wave economic phenomenon, tied to overall long-wave fluctuations in economic growth of roughly a quarter of a century. Since it is by now generally accepted that the Western industrialized economies are well into a period of very slow growth likely to last to the end of this century, following a two and a half decade period of very rapid growth, Freeman concludes that "in the race between job-generating investment and technical change and job-displacing technical change, I would expect job displacement to draw ahead in the 1980s."[16]

Others would argue that job displacement will not merely draw ahead but leap ahead in the 1980s. Consider these few examples from the decade of the 1970s in industries already significantly affected by the introduction of more advanced electronic technology:

— In making television sets in Britain, the number of man hours required dropped from 100 in 1973 to 50 in 1978 (and dropped still more in Japan).
— Workers making cash registers in a British firm declined from 2000 in 1977 to 1350 in 1979 while the number of moving components dropped from 250 to less than 25; the corresponding figures for a large North American manufacturer were 37,000 in 1970 and 18,000 in 1975.
— Employment in the telecommunications industry in the United Kingdom dropped from 88,000 in 1974 to 65,000 in 1978, while a major U.S. manufacturer halved its work force from 39,200 to 19,000 in six years (1970–1976).[17]

As we have noted, the most substantial impact on employment through the introduction of micro-electronic technology is expected to occur not in manufacturing from which these examples are drawn, but in the information sector. One estimate suggests that, although the volume of banking services will continue to grow, employment in French banks will decline 30 percent in the next ten years through application of automated information processing.[18]

Micro-electronics is not the only variable affecting the character and performance of industrialized economies. Even though they lie outside the scope of this discussion, the existence of other variables should be noted, lest the conclusion be reached that all of the possible changes in these economies in the 1980s will be attributable to the silicon chip. At least two other imponderables seem worthy of mention in this context. Both may well have — some would argue, already are having — a profound impact on the overall level of economic activity in industrialized societies, as well as on the nature of technological change and employment. But, because these imponderables are relatively recent phenomena as significant economic variables and have exhibited highly dynamic, if not volatile, characters, especially the latter, their impact on economic performance, technical change, and employment is extremely difficult to analyze with any precision in the recent past, let alone forecast in any meaningful way into the future. I refer to the environment (and policy interventions to protect/improve it) and energy (or more broadly, non-renewable natural resources, of which energy is surely at the moment the most important).

Both phenomena have generated a voluminous literature on their recent and prospective economic impact, much of it polemical. But even where there is systematic analysis, conflicting conclusions are often reached. Regardless of the character of the analysis of these phenomena, however, there is some growing evidence that they are likely to have a significant impact on the nature of technological choice in industrialized societies in the future. Analysts and advocates of technological alternatives in environment and energy, such as Amory Lovins with his "Soft Energy Paths" and Barry Commoner with his "Solar Transition," who were dismissed by mainstream economic and political decision makers a few years ago, are increasingly getting a more serious audience as their earlier predictions and prognostications are proving to be close to the mark of subsequent experience.[19]

All of these factors — in the long-term, cumulative impact of major technological changes which started in the 1960s, especially the micro-electronic revolution, and such variables as energy and the environment in the 1970s — coupled with the long-wave character of overall economic performance, add weight to the paradigm-shift thesis in forecasting the likely impact of technological change on employment in industrialized societies during the remaining decades of this century, if not into the early decades of the next.

Alternative Responses to a "Paradigm-Shift" in Technological Change and Employment

Given the conflicting results of analyses of these phenomena and the controversy associated with them, it is not surprising that there is wide variation in the prescriptions offered for what should be done to respond to the problem of technological unemployment in industrialized countries, if indeed there is a problem.

At one pole are the "business-as-usual" and "history-will-repeat-itself" advocates who would argue that we need to continue to follow existing "hard energy" paths as the surest way to reinvigorate overall economic growth and thus to expand employment. We should not give undue weight to environmental protection, which may slow down growth rates and have an adverse impact on employment, and press full steam ahead with technological efforts to increase labor productivity, even if this means short-term job displacement. Much of the problem of technological unemployment, in the view of some, can be solved simply by shortening the work week, a step which should surely be possible without diminishing our standard of living if technological innovation has been able to boost labor productivity sufficiently. An economist of no less stature than Nobel Laureate Wassily Leontieff has recently taken such a position.[20]

Little attention appears thus far to have been given by advocates of this general persuasion to the formidable political problem of intersectoral distribution of the "automation dividend".[21] The significance and severity of this issue will vary widely in terms of the character of the economy and nature of organization of the work force. In Sweden, with a substantial proportion of the work force organized and with broad-based national trade union organizations, the intersectoral distribution problem, while certainly difficult, is likely to be less traumatic than in some other industrialized countries and will be accomplished through tax-induced income transfers in the form of widespread public services, including public service employment for those who might otherwise be displaced from the job market through automation. Indeed, these practices already are being followed to a considerable degree and are a major reason why Sweden, alone among the OECD countries, shows a lower unemployment rate in 1977 than the 1962–73 average; according to Table 9.2.[22]

The problem will be very different in the United States, where a minority of the workers are unionized and there is no effective, broad-based national trade union organization. Here, the political battles between management and labor and between the organized and unorganized segments of the work force on how to divide the spoils will be fierce. In fact to the extent there is a big "automation dividend" in the United States, the political strug-

TABLE 9.2   Levels of Unemployment in OECD Countries
(Percent of Labor Force)

	1962–73 average	1973	1974	1975	1976	1977 (latest 3 months)[1]
Canada	5.3	5.7	5.4	7.1	7.2	8.1
U.S.A.	4.9	4.7	5.6	8.5	7.7	7.1
Japan	1.3	1.4	1.4	1.9	2.0	1.9
Australia	1.6	1.9	2.3	4.4	4.4	5.1
Belgium	2.1	2.8	2.6	4.5	5.8	6.4
Denmark	· · ·	1.1	2.5	6.0	6.1	7.0
Finland	2.4	2.3	1.7	2.2	4.0	5.5
France	1.8	2.2	2.3	4.0	4.2	4.6
German F.R.	1.3	1.3	2.6	4.8	4.7	4.4
Italy	3.6	3.8	2.9	3.4	3.9	6.8[2]
Netherlands	1.4	1.0	3.3	4.7	5.1	4.7
Norway	0.9	0.8	0.6	1.2	1.1	0.9
Sweden	2.1	2.5	2.0	1.6	1.6	1.6
Spain	· · ·	2.3	3.2	3.8	4.9	5.4
U.K.	2.4	2.4	2.5	4.2	5.8	6.9

Notes: National definitions, not adjusted for international comparability.
1. Seasonally adjusted (except Italy and Spain), mainly second quarter.
2. New survey definitions. Estimated figure on old definitions was 4.1 percent.

Source: C. Freeman, Paper presented at the Conference on "Science, Technology and Public Policy: an International Perspective," University of New South Wales, December 1977 (Based on OECD, *Economic Outlook* and *Selected Economic Indicators.*)

gle over its distribution is likely to rival that of the Third World market economy countries, which will face increasingly severe intersectoral distribution problems in the 1980s if present trends continue to widen disparities between the modern industrial and traditional, largely rural sectors of the economy.

Of course, there may not be any "automation dividend" — or at least not one of very great size if those of a different persuasion are closer to the mark in their projections of what will happen if a "paradigm-shift" actually is seen to be occurring on a substantial scale. For compelling political and social reasons, a series of policy interventions and changes may be made to try to alter the direction of technological change toward greater job generation while responding to environmental and natural resource constraints. Expressed in concrete terms, here are four interventions suggested by Christopher Freeman to achieve such a goal in the 1980s:

1. Greater support for innovative small firms.
2. More serious attention to alternative labor-intensive techniques in major areas of public investment which are particularly capital-intensive, such as energy.

3. Support for capital-saving R & D projects in public laboratories, development contracts, and government procurement contracts.
4. Re-examination of tax and other incentives which give a bias to the choice of capital-intensive technologies in private industry.[23]

It should be emphasized that if the situation turns out to be as serious as some analysts predict, the foregoing propositions will constitute hardly more than the opening wedge of an interventionist response. Just how serious the situation is likely to be is the subject to which we now turn our attention.

Chips and Jobs: Weighing the Balance

For those who argue that micro-electronics will have widespread disruptive impact on industrialized economies, two difficulties arise — one historical and the other analytical. More than once in the past, and particularly within recent memory (the "automation scare" of the late 1950s and early 1960s), doom criers have predicted the collapse of economic systems or something akin to it — and it has not happened.[24] Will history repeat itself? No one knows, of course. Sometimes it does and sometimes it does not. But the "Cry Wolf" phenomenon adversely affects the credibility of those who argue that industrialized economies will have a hard time of it in the 1980s, adjusting to the pervasive and often radical changes in how we produce goods and services as a consequence of the widespread diffusion of micro-electronics.

The analytical difficulty arises from the extremely complex, interactive, and dynamic character of economic variables in industrially advanced countries. The fact is that no one can prove the case conclusively for or against extensive unemployment being caused in the future by micro-electronic technology. Virtually all serious analysts and observers of the ecnomic scene in the industrialized countries would agree that things will change — and indeed are changing. The critical questions are how rapidly, how much, and with what overall consequences in terms of employment.

There is, furthermore, the possibility that if sufficiently forceful and persuasive cries are raised now, governments will respond with policy interventions to ameliorate the impact of micro-electronic technology. This will enable critics of the doom criers to argue, five or ten years from now, that once again the doom criers were wrong and their critics right because the dire changes did not materialize.

As the European Trade Union Institute has noted in its study of the impact of micro-electronics on employment, labor-saving innovation does not necessarily have an adverse impact on employment. Technological unemployment can be said to occur when output per head rises faster than total output and is not offset by other variables such as reduction in working time. In the Federal Republic of Germany from 1971 to 1975, these conditions obtained (despite a decrease in hours worked) so that employment fell by 1 percent and the unemployment rate increased from 0.7 percent to 4.6 percent.[25]

The problem of getting a fix on the technology factor in unemployment is compounded by the circumstance that macro-level unemployment data are not ordinarily classified and analyzed with technology as a key variable. Rather, unemployment is usually identified as having any one of three characteristics — "structural unemployment" (where workers displaced by technology are not re-employed due to lack of appropriate skills or geographical or sectoral immobility, "capital shortage unemployment" (insufficient capital of the proper type to employ the displaced workers in making goods or providing services for which there is a demand), and "demand-deficient unemployment" (insufficiency of demand in relation to the potential level of output).

Given these circumstances, not to mention other factors such as variable elasticity of demand in relation to prices of new products, or services, or changing capital, or labor intensities because of technological innovation, those critics of exponents of technological unemployment introduce an unrealistic and analytically unattainable demand (at least at any credible qualitative level) when they insist that such exponents produce hard numbers on *future* unemployment identifiably due to technological change. But, it is possible to identify recent changes in employment at the individual enterprise and sectoral levels where technology has played a key role. In addition to the examples already given from manufacturing, other examples can be adduced from other sectors of economic activity such as banking and insurance (where rates of employment growth have dropped sharply as volume and types of services have increased in the last five years); office work (the local government authority in Bradford in the United Kingdom reduced its typing staff from 39 to 19 with the same volume of work after introducing word processing machines, and a Norwegian insurance company uses 13 persons to do what 75 did two decades ago with introduction of the same equipment); maintenance (a Canadian telephone exchange has reduced its maintenance staff from 11.0 to 1.7 men per 10,000 lines with installation of an electronic exchange); and even the distribution sector (although other factors are also involved, technology has played a significant part in reducing employment in retailing in West Germany by 5.9 percent between 1972 and 1977, 12.5 percent in Italy from 1975 to 1977, and 2.7 percent in Belgium over the same period).[26]

The outlook for the 1980s in the industrialized countries is for an intensification of these trends without offsetting increases in employment elsewhere in their economies. Several factors point in this direction. One concerns the often dramatic reduction in employment which occurs with the introduction of technology based on micro-electronics. Thus, at Volvo's Torslanda plant near Gothenburg in Sweden, an automated welding line has reduced the number of workers from 100 to 20 over two shifts. A fully computerized production system covering both body construction and finishing, now being discussed at this plant, would reduce employment from 1030 to 60.[27]

Another factor is the changing structure of production, with the forward movement of components into the final assembly stage. A mechanical watch takes some 1,000 operations to assemble compared with the assembly of five components in an electronic watch. One consequence: a reduction of employment by 46,000 in the mid-1970s in the Swiss watch industry. It appears unlikely that there will be offsetting increases in employment in the component-manufacturing and capital-goods industries as a result of this shift, and the increasing use of automated machinery such as industrial robots because of the use of micro-electronics for both design and production control systems (e.g., in the machine-tool industry already mentioned). Texas Instruments, by way of illustration, has stated that the world demand for 64K RAM (Random Access Memory) devices in 1985 can be met with 1,000 employees.[28]

Much of the discussion thus far has focused on the labor-saving character of micro-electronic technology. But the micro-electronic technology can be capital-saving as well — for example, through optimization of output from production facilities through increasingly more sophisticated control systems or more extensive use of production equipment with marginal increments in the wage bill (an automated factory can be run around the clock with only a handful of workers involved). Indeed, Ray Curnow, co-author with Ian Barron of a major work on the information sector in the British economy, goes further and argues that since less material and energy inputs and moving parts are involved in information technology goods or their use, there will be a structural shrinkage in the economy as the importance of the information sector increases through the diffusion of micro-electronics.[29] If such a structural shrinkage occurs, it will hardly make any easier the task of generating productive and meaningful work in industrialized societies in the 1980s.

These factors are compounded by the slow rate of growth for the industrialized economies generally forecast for the 1980s, or at least the first half of the decade. But this will not necessarily mean a deceleration in the rate of introduction of new technologies, at least in the case of Japan and Western Europe. The vital importance of maintaining or improving international competitiveness for these economies will lead to the continuing introduction of micro-electronic technology in order to achieve cost reductions essential to maintaining competitiveness. In the meantime, employment expansion in the tertiary sector of these economies is beginning to slow down or stop, and, as employment is reduced in other sectors of economic activity, overall employment rates have been rising. In the European Community, registered unemployment rose from 3 million or 2.9 percent in 1974 to 6 million or 5.5 percent in August 1979. Youth unemployment has grown especially rapidly. The unemployment rate among those under 25 has risen from 2.9 percent in Italy, and from 2.9 to 14.1 percent in the United Kingdom.[30] A similar pattern has been occurring in the United States, especially among racial minorities (with unemployment rates as high as 45 percent in New York City for young Blacks).

One forecast indicates that unemployment in Western Europe as a whole will increase to over 10 million or 7.5 percent by 1983 and 12.5 million or 9.1 percent by 1990. But this forecast is based on two assumptions, both of which seem unlikely — that there will be an average annual 3 percent growth in GNP until 1990 and that displacement of workers through micro-electronic technology in the service sector of the economy will be offset by equivalent employment growth in the electronic data processing sector.[31] In other words, the position is probably going to be considerably worse than this glum forecast, based as it is on two optimistic assumptions, suggests.

What all this adds up to for the industrialized countries in the 1980s is the prospect of a continuing decline in industrial employment, and a rapid reduction in the expansion of the service sector (if not an outright decline in some countries), even if there should be increased output and the introduction of new products and services. Such a statement must always be accompanied with the qualification that rather drastic public policy interventions could change the course of these trends, or at least moderate or disguise them (through, for example, greatly expanded public sector unemployment).[32] But in the absence of such intervention, the creation of a few thousand jobs for design engineers and computer programmers through the diffusion of micro-electronics will not offset the loss of hundreds of thousands of jobs for clerks, typists, and factory workers.

Typically, in the past it has been workers who are the first group to pay the social costs of technological change. The fact that in industrialized societies, much of the expressed concern over the impact of micro-electronic technology has come from workers and the trade unions which represent them is a circumstance not to be dismissed lightly. Their claims may be strident at times, but if they are, it is for a purpose: to make certain their voices are heard and heard and responses initiated before the full brunt of this revolution is on their backs.[33]

In the long run, of course, assuming intersectoral income distribution problems and a host of other questions such as social control of technological change and creation of meaningful alternative forms of productive activity can be sorted out, the silicon chip with the dramatic increases in productivity and the perhaps less spectacular but also significant efficiencies in energy and raw material use it can make possible offers great promise for improving human circumstances in the industrially advanced countries.[34] Its impact on the Third World is likely to be a much different proposition, and to that set of issues the discussion now moves.

Technology and Employment in the Third World: A Hobson's Choice

Whatever may be the magnitude and character of labor displacement from technological change in industrialized countries, it pales almost to the point of insignificance in relation to the problem faced by the developing countries of providing meaningful and productive work for most of their citizens. By most measures of the impact of economic performance on individuals, these countries face an equity crisis of arresting proportions in the 1980s if such a crisis it not already upon them.

Developing countries loosely lumped together under the rubric of the "Third World" are, in fact, enormously diverse in physical size, level of economic development, and rate and character of individual participation in and benefit from the economic system. Given these circumstances, it is better to use actual examples. I shall use three to make different points about the changing political economy of the Third World and the impact of technology on it — India, China, and Brazil. All three are important for reasons of size: the first two between them contain more than a third of the population of the entire world and half that of the Third World. Brazil, while much smaller in population, is large in physical area and is a country of major significance in Latin America. All three are among the more advanced or "newly industrializing" Third World countries or NICs. All three are faced with complicated problems in terms of individual participation in the economic system, whether in income distribution or employment or both.

The equity crisis in the Third World has three principal components — acute material deprivation, highly unequal income distribution, and very substantial unemployment, underemployment, and employment at very marginal rates of productivity. In Brazil in the 1960s, for example, while GNP was growing at the rate of 5 to 6 percent a year and Brazil was aggressively importing capital intensive technology to develop its modern industrial sector, the share of the national income going to the top 5 percent of the population rose, according to official government statistics from 27 percent to 39 percent, while some private estimates place the latter figure at 46 percent.[35] In India, while estimates vary, some 40 percent of the entire population (and more than 50 percent of the rural population) are generally believed to be below the Indian "poverty line," defined in spartan terms as income sufficient to provide for enough calories to stay alive.[36] Even China, while it is usually thought to have eliminated the most acute forms of material deprivation, which afflict substantial proportion of the Indian and Brazilian populations, is still, as the Chinese themselves point out, a poor country with widespread poverty, or at least large numbers of persons living with very modest material means.[37]

The question of employment is a complicated one in these countries. There is no convenient single measure that reveals the magnitude of the problem such as the number of persons registered on unemployment rolls. If that were the case, China would presumably show virtually no unemployment, and yet China clearly has a massive problem of coping with large-scale employment at very marginal rates of productivity in the rural areas. More data are available on India, but even here, the situation is complicated as Pramit Chaudhuri has shown in a recent paper. There are some indicators — for example, the number of landless agricultural laborers for whom unemployment and underemployment are particularly acute, increased by 90 percent in the 1960s. It is estimated that the number of Indians unemployed, underemployed, or employed at very marginal rates of productivity is between 120 and 200 million, mostly in the rural areas (since that is where 80 per cent of the population is).[38]

But, what lies ahead for India is far worse. According to K.N. Raj, the male labor force alone — i.e., only half the population — is growing in the decade 1975–1984 from 152 million to 196 million.[39] Thus, even assuming that the present population is already gainfully and productively employed, 44 million new jobs would need to be created during the decade now half over.

To show the irrelevance of the modern industrial sector in meeting employment needs of this magnitude, private industrial employment in India — which, like Brazil, has a dynamic modern industrial sector — has fluctuated between 6.5 and 6.8 million in the ten-year period 1967–1976.[40]

Third World countries like Brazil, China, and India are indeed confronted with a Hobson's choice in devising strategies to cope with the task of providing meaningful and productive work for their peoples. Completely implausible increases in manufacturing output would be required to absorb current and future population growth through industrial employment, let alone the huge backlog of unemployed, underemployed, and marginally employed. For India, the *annual* increase in industrial employment would need to be 22.1 percent, and for Brazil, 16.2 percent.[41] China, apparently recognizing the hopelessness of the problem, is working on a three-sector national development strategy for the remaining decades of this century: a modern, urban-based industrial sector (for which strenuous efforts are now being made to acquire the latest technology from abroad), a high productivity agricultural sector (mostly in the form of state farms surrounding major urban areas) to serve the modern industrial sector, and the mass of the people remaining in low-productivity agriculture and small-scale rural industry, striving to be as self-sufficient as possible.[42]

The Hobson's choice becomes even more glum if the labor-absorption possibilities of agriculture are realistically considered. Pramit Chaudhuri's analysis of Indian agriculture suggests that under the most favorable crop-ping and technology mix in terms of employment generation, estimates run as high as 50 percent of the rural population, which would be required to find some form of non-farm employment if subsistence income levels are to be achieved.[43]

People vs. Machines: The Relevance of "Appropriate" Technology for Developing Countries

Given these melancholy circumstances, what role is there for so-called ap-propriate technology — i.e., technology that is more labor-using and capital-saving rather than the other way around, as is characteristic of industrial technology now in use in the economically more advanced countries? It is always better to light even one candle than to curse the darkness. Ap-propriate technology, therefore, certainly has a place in the scheme of things in the Third World as it grapples with its massive equity crisis.

But it is clearly no panacea, and overemphasis on labor-intensive technology with low productivity could become a trap for poor countries, rendering still more difficult their efforts to meet the most essential material needs and provide meaningful and productive work for their populations. It is no solution for the quantitative, politico-economic reasons already given. Changes in technology as such cannot possibly cope with absorption of the rapidly growing labor forces in countries like Brazil, China, and India. Nothing short of a social revolution will do that, and even then the results cannot be guaranteed.[44] These countries, furthermore, need the much higher productivity of modern technology even more than the industrially advanced countries, if they are ever to find a way out of their present predicament.

But labor-intensive technology is also no solution for qualitative reasons. For producing certain kinds of goods and services, machines are simply better than human beings, and in some cases (ability to withstand high temperatures in metal fabrication, for example), the only means. Machines have greater speed, accuracy, and reliability than people, all being qualitative characteristics that contribute to productivity. Thus, for techni-cal, as well as productivity, reasons, developing countries must have modern technology, even though it is ordinarily more labor-saving rather than labor-using and, thus, seemingly exacerbates the problem of absorbing large unemployed, underemployed, and rapidly growing labor forces in these countries.[45]

Micro-electronic Technology and the Third World: Resisting the Luddite Trap

Even if so-called appropriate technology is not the total solution to the employment problems of the Third World, common sense compels us to question the relevance of a form of modern technology such as micro-electronic technology that so dramatically increases labor productivity. For a curious set of partly perverse reasons, micro-electronic technology does have a significant potential role in helping poor societies grapple with their equity problems. It could also help them to achieve a more autonomous, less dependent role in the international system, although it may not, certainly in the short term. Since I recognize that these assertions tax the indulgence of most readers, I shall endeavor to argue the position as persuasively as possible. It should be understood that the case is largely speculative because of the embryonic state of our present knowledge of the economic, social, and political consequences of widespread adoption of micro-electronic technology, as was pointed out at the very beginning of this study. I also perceive the short- and medium-term outlook for North-South relations quite differently, so that readers who think I have taken leave of my senses in the next few paragraphs should persevere to the end.

First and perhaps foremost is the hypothesis that micro-electronic technology would achieve significant gains in the productivity of *both* capital and labor. If it is not possible to base future development strategies primarily or exclusively on labor-intensive "appropriate" technology for both the quantitative and qualitative reasons just advanced, it is certainly an advantage to capital-poor developing countries to get greater output per unit of capital invested than not, especially if the productivity gains become substantial as micro-electronic technology shows *promise* (not yet much actual performance) of making possible.

Other characteristics of micro-electronic technology are also relevant to the needs of developing economies — for example, its relative flexibility and the possibilities it offers for greater product differentiation, which in principle mean that developing countries can take a basic technology and adapt it more easily to suit their own social requirements or raw-material availabilities. In a similar vein are the possibilities for decentralization of production while maintaining high levels of standardization, which may enhance the possibilities for developing more rapidly substantial, small scale, high-productivity industrial sectors in the Third World and diminishing transportation costs and urban concentrations (although it must be noted that it will take a long time in most developing countries to create the economic and educational infrastructure in the countryside so that modern micro-electronic technology can be effectively used. It should be underscored that these are theoretical possibilities for the longer term, not immediately implementable alternatives.

It is clear that the industrialized countries will far more readily capture the benefits of micro-electronic technology, in no small measure for these infrastructural reasons. Micro-electronic technology may be *relatively* more accessible to developing countries than other forms of advanced technology, at least as far as the hardware is concerned — that is to say, they should be able to buy it (at a price), even if they do not make it.[46] In the case of Britain, as Ray Curnow has pointed out, "from an economic point of view, it is far more important that the UK use the new technology than that the UK should provide it."[47] The same may be true for some Third World countries.

It will certainly be the case for some time that manufacture of the basic hardware in the form of integrated circuits on silicon chips will be concentrated in one or more of the major industrialized countries, but sources of supply, given the enormous stakes involved, are likely to spread pretty rapidly. But, the real possibilities and challenges of micro-electronic technology lie in the software applications, and here, developing countries could get substantial access under relatively non-dependency perpetuating conditions (i.e., in disembodied form without entangling and enduring investment and licensing arrangements) — *if* they prepare themselves and are willing to exert themselves. The key to access here lies in the training of large numbers of design engineers and computer programmers. Some of the newly industrializing countries (NIC) have already made a modest beginning in this direction, but will have to do *far* more before the conditions suggested on the preceding sentence will obtain.[48]

Obviously, the economic and social realities of developing countries cannot be ignored, and conditions there are often not the same as in industrialized countries with very high labor costs. Therefore, the circumstances under which it will make sense to use micro-electronic technology in developing countries will differ greatly from the industrialized countries, even from one developing country to another. The danger is that economic and political elites in the Third World, having been effectively seduced by hardware and software salesmen from the North, will become infatuated with micro-electronic technology, seeking its indiscriminate application.

Indeed, we have thus far been proceeding as though micro-electronic technology, while economically beneficial though socially disruptive, is politically benign. Nothing could be further from reality. Certain applications of micro-electronic technology offer distressing means of enhancing the effectiveness of technologies of destruction and repression. Therefore, the question of who controls critical elements in this technological revolution becomes crucial, among, as well as within, countries.

Because there is no known way of getting this technological genie back in the bottle, wholly or selectively, the best strategy to minimize the possibilities for exploitation and control is to diffuse it through as many channels and as widely as possible. In fact, there is an important reason for Third-World countries to get in on the micro-electronic act. Failure to do so will cast them in a more dependent relationship on the First and Second Worlds. The industrialized countries will find it all the more tempting to use the advantages of micro-electronic technology in speed, flexibility, cost, reliability, and control to maintain their presently dominant position in the international economic and political system. They will sell the products made possible by micro-electronic technology and their services in applying it at inflated prices, unless Third World countries become effective bargainers by developing their basic competence in using and understanding this rapidly developing arena of economic activity.

Impact of the Micro-electronic Revolution on North-South Trade and Investment Flows

While any such broad generalization is obviously subject to exception, one reason for Third World adoption of micro-electronic technology is *not* to gain or enlarge access to industrialized country markets for Third World manufactured goods. Over and over again the cry comes from Southern spokesmen for the New International Economic Order (and from Northern Spokesmen who say they wish the South well, a recent example being Robert McNamara of the World Bank at UNCTAD V in Manila in May 1979) that Northern countries must open their markets to Southern developing country manufactured goods if the latter countries are ever going to find a way out of the poverty and deprivation that afflicts their peoples.[49] There are two things wrong with this cry: It is not going to be heard and, even if it were, it would not yield the result contemplated.

We have already examined the second reason in looking at the Hobson's choice confronting the Third World. For most Third World countries, certainly the larger ones like Brazil, India, and China, a buoyant export-oriented modern industrial sector aimed at penetrating industrialized country markets with its output, no matter how dynamic, is not going to solve the employment or equity problems in these economies. (The island city-states like Singapore and Hong Kong or the NICs on the periphery of the Eurasian landmass such as Taiwan and Korea may be partial exceptions, but their combined population is minor in relation to the continental giants such as India and China.) In fact, substantial evidence suggests that this kind of export-led strategy of national economic development exacerbates the employment and equity problems by unduly misallocating scarce capital resources to the modern sector instead of to the deprived rural sector and by diverting productive energies and raw materials from meeting the material needs of the local populations to fulfill the seemingly insatiable appetites of consumers in rich countries with what are often economically marginal, socially trivial goods.[50]

It is a fair question as to the extent of concern of Third World political and economic elites with problems of equity and employment.[51] But they appear no less concerned, at least at the rhetorical level, than their counterparts in the First and Second Worlds. They are interested, as are their counterparts elsewhere, in maintaining political stability in order to retain their power, and to the degree that issues such as employment and equity have no impact on stability, they are doubtless concerned with them as well.

Efforts to maximize Southern penetration of Northern markets with manufactured goods are more likely to compromise Southern efforts to achieve a more autonomous and less dependent role in the international system. The export-oriented modern industrial sectors of Third World economies will continue to depend on technology imports from the industrialized countries because typically these countries are the only source of economically competitive technologies that will produce what industrialized country consumers will buy. The increasingly interlocking, but still dependent North-South relationship that will result from such a Southern strategy of development will make Southern countries exposed to Northern pressure on a continuing basis by laying down political and economic conditions, as the price of continued access to those markets. Weaker Southern political economies will find it hard to resist this pressure in any show of strength with their Northern adversaries.

This is not interdependence, as Northern advocates of such a relationship would have us believe. True interdependence requires two roughly equal partners. This is simply a more intimate pattern of the prevailing dominance/dependence relationship that now characterizes the global political economy, a conclusion strikingly demonstrated by Graciela Chichilnisky in her critical analysis of the second generation world models (the Fundacion Bariloche's Third World-oriented model, *Castastrophe or New Society;* the UN-commissioned model on *The Future of the World Economy* by Wassily Leontieff and others; and the RIO group in the Netherlands headed by Jan Tinbergen, *RIO: Reshaping the International Order).*[52] Even the much-vaunted countervailing Southern bargaining chip of raw materials, on which there is more below, is likely to be short-lived and more illusory than real.

The second reason why this cry is wrong is probably more decisive. It is simply not going to be heard. By the time the full brunt of the micro-electronic revolution hits Northern economies, they will be far more protectionist than they are now, striving to save every job they can from "cheap labor" competition through imports. All bets are off on the liberal trade rhetoric of industrialized country political elites when they are confronted with unemployment rates of 10 to 20 percent at home.

In any event, developing country expectations about industrialized country markets have already reached totally unrealistic levels. If all the shoe manufacturers in Asia were actually to increase exports to Europe to their projected levels by 1980, European domestic shoe production would have to stop and per capita shoe consumption would need to increase by several hundred percent.[53]

As the introduction of micro-electronic technology gathers momentum in industrialized countries, with its promise of significant, possibly dramatic, increases in labor productivity, the relative comparative advantage of Third World countries in more labor intensive production (e.g., traditional manufactured goods like textiles and shoes) will begin to be eroded. The result will be increasing difficulty of access for Southern manufacturers on the basis of cost competitiveness and a probable withdrawal of Northern multinational corporation investment and production facilities from Southern countries on grounds of decreasing profitability and minimization of risk, the latter factor being paradoxically enhanced as Southern countries become more adept and assertive in investment and technology bargaining.

At least one large multinational has allegedly already concluded that it is better corporate policy, and apparently no less profitable, to build automated factories in Europe in the future rather than continue labor-intensive electronic manufacturing in Asia.[54] And with automated looms, some European textiles are again becoming cost-competitive with those made in the Third World. Both are signs of times to come.

In short, the issue of difficulties for Third World market penetration of the industrialized economies does not depend on rising unemployment in these economies due to technological change, admittedly a debatable proposition, as I have indicated in my earlier discussion of the question. Of course, rising unemployment will surely aggravate the problem of Southern market access, and indeed, the present and persisting unemployment levels in OECD countries are already threatening to set off a protectionist response as the recent report of the OECD Secretary General on the NICs implicitly recognizes.[55] Shifting comparative advantage, based in significant measure on the introduction of micro-electronics with its substantial changes in labor productivity, will enable the Northern economies to recapture many of the traditional or "twilight" industries such as shoes, textiles, and garments, which were beginning to move southward.[56]

There is another hypothesis worth mentioning in the context of future patterns of North-South trade and investment, but it is hardly more favorable to Southern interests. Indeed, it may simply mean an enlarged scope for exploitation, especially through the medium of internal transfer pricing by Northern-headquartered MNCs. Trade barriers erected by developed countries as a political consequence of structural unemployment caused by introduction of micro-electronic technology, may well fall more heavily on developing country-owned firms, because subsidiaries of MNCs will be able to escape the brunt of these barriers through intra-firm trade (i.e., the products, only some components of which will be made in the Third World, will not be classed as imports). But, in order to maintain competitiveness with industrialized-country manufacturers taking advantage of the greater productivity of capital and labor through micro-electronic technology, MNCs will need to underprice their Third World-manufactured components through the device of internal transfer pricing, thereby shortchanging Third World subsidiaries (or joint ventures) and their workers.[57]

In today's international political economy, Southern countries are generally regarded as having a good and increasingly substantial bargaining chip in their control of raw materials needed by the industrialized countries to maintain their affluent living standards and keep their economies functioning at a satisfactory level. There is no doubt about Northern dependence on the South in the short terms, oil being the most obvious example. But this bargaining chip is likely to be short-lived and not nearly as potent as might otherwise be thought, certainly if we look at trends in the global political economy in decades rather than years. The revolution in materials technology, while more futuristic than the micro-electronic revolution, is not that far away if some of Barry Commoner's arguments about the solar transition have any validity.[58] Paradoxically, that revolution in materials technology, with its thrust toward renewable bio-mass resources, may be accelerated to the extent Southern countries try to use this bargaining chip. There seems to be little doubt that OPEC greatly accelerated the solar transition in Northern countries with its price revolution in petroleum, beginning in 1973.

All of these trends and forces are likely to add up to less, rather than more, economic integration of North and South in the remaining decades of this century. Thrown increasingly back on their own resources and initiative, Southern countries may well begin to expand trade, investment, and technology links among themselves. This is, in fact, already beginning to occur as some of the more advanced developing countries are becoming technology exporters to other Third World countries.[59] Whether this will necessarily lead to a less exploitative international system, let alone achieve greater equity within Third World countries, remains to be seen.[60] On the face of it, economically and politically weaker Third World countries would seem to be as much at the mercy of their strong, more highly developed Third World brothers as they are with industrialized countries. On the other hand, they may be no worse off than they are now in a highly dependent relationship with the industrialized countries. In fact, if they are adroit, they may be able to play different groups of dominant actors off against one another.

Countervailing Views on North-South Relations

There is always, of course, the possibility that the modern industrial sectors of some Third World countries, especially the export-oriented NICs, will seek to capture, retain, or recapture the same comparative advantage made possible by sharp changes in productivity through micro-electronic technology, just as the industrialized countries are now beginning to do. Indeed, some NICs are already starting to move in that direction, with Korea as, perhaps, the most notable example. Certainly the same pressures to do so, in order to maintain international competitiveness, exist for Korea as they do for the OECD countries. With industrial wages now half those of Japan, Korea has only one-third of Japan's productivity.[61]

But efforts along these lines by the NICs, at least thus far, have not borne much fruit. Korea is seeking to upgrade its electronic technology through a specially created institute that has set up a branch in Silicon Valley in California, but the U.S. electronics industry, especially the semiconductor industry, has been less than forthcoming in providing access to advanced technology (as distinct from products or components used in assembling products).[62]

There is also the view, argued persuasively by J.B. Donges and James Reidel of the Institut für Weltwirtschaft in Kiel, that the major reason for limited NICs' penetration of industrialized country markets is not because of barriers, tariff or non-tariff, created by the industrialized countries, but because of limited supplies of exportable products by the NICs.[63] If this view is correct, it may lead to expanding North-South trade in the short term, provided the NICs are able to overcome their constraints on supply and the industrialized countries do not increase trade barriers as the OCED Secretary General clearly fears they will.[64]) But, if the thesis of shifting comparative advantage through the application of micro-electronics in the traditional or "twilight" industries is anywhere near the mark, removal of supply constraints by the NICs on exports to the North will only make their adjustment problems more difficult a few years later.

Yet another countervailing view is that the Northern need for markets in the South will make them more accommodating about giving the NICs access to their domestic markets in order that the NICs can earn the foreign exchange needed to purchase Northern manufactured goods, especially capital and infrastructural goods. The character of many of these goods is likely to be substantially changed through micro-electronics in the 1980's. These developments will affect both the NICs and the industrialized countries. The former will want to buy them in order to remain competitive, the latter to sell them in order to realize economies of scale or spread large overheads, especially for R & D, over more units of production.

There is, of course, merit to this view, but I am inclined to think that on balance, to the degree that the Northern need for markets is that strong, a more likely strategy from the Northern point of view than opening up their markets to the NICs, will be to finance capital goods and infrastructure investments through credits and tied aid. These instrumentalities have the twin virtues of maintaining greater Northern control over the technology the South acquires and perpetuating a dependency relationship through continued Southern indebtedness to Northern governments and financial institutions. Of course, these are not absolute alternatives, and some of each will occur in the future as they have in the past. The issue is more one of degree.[65] What can be said at present is that, even though Southern purchases of Northern industrial products have by and large been expanding faster than NICs' exports to the North, both imports and exports are, at least for the OECD area as a whole, at very small, almost trivial levels. The really massive growth in international trade in the postwar era has been among the OECD countries themselves.[66] A not impossible view of the future global political economy, therefore, is one in which the OECD area remains largely self-contained, using micro-electronic and other technological advances to regain whatever comparative advantage it may have lost to the NICs in the 1960s and 1970s, while it increasingly exploits new advances in materials technology, especially biotechnology, to diminish its remaining dependence on Southern raw materials.

Such a global political economy may not be best for certain interests in the Third World, even for a large proportion of their populations, as some observers of the changing character of North-South relations would argue. For the South, given their far weaker, more dependent position in the international system, I would submit that the question is not so much what it best, as what is least worst. But, regardless of what is best or least worst from a Southern view point, future patterns of North-South relations are likely to be primarily determined by the particular constellation of political and economic forces and technological trends emerging within the North, especially the OECD countries.

The Micro-Electronic Revolution and the Future of the Global Political Economy

All of this adds up to a quite different picture from that usually painted by strident advocates for the New International Economic Order or apologetic Northern well-wishers and supporters of NIEO. We may end up with a new global political economy all right, but one rather different from that now imagined by these advocates and apologists. Can one little silicon chip — or even lots of little silicon chips — actually rearrange the political and economic landscape of the world along the lines I have been suggesting? The answer, of course, is no. It is not the chip that does the rearranging, but political and economic actors who use the chip in certain ways. The chip is a tool in their hands to be used for good or for ill, although it is the unusual technological properties man has managed to inscribe on the chip that give the tool such efficacy.

Micro-electronic technology offers the promise of remarkable, perhaps extraordinary, increases in the productivity of capital and labor and in more efficient use of raw materials. For no group of countries in the modern world is this more important than the developing countries. Unless they can increase their productivity dramatically and use their resources more efficiently, there is no way out of the poverty that engulfs them. And, for industrialized countries with major unmet social needs, these technological changes also present the opportunity of meeting those needs, as well as constructing less exploitative relationships with developing countries. But for both groups of countries, the most vexing issues lie not in applying new technology to increase productivity, but in grappling with the political, social, and economic problems of intersectoral income distribution and creating vast numbers of new opportunities for meaningful and productive work while searching for alternatives to the historical link between personal income and productivity.

The major task for developing countries is to harness the potential of these revolutionary technological changes to increase productivity of the modern industrial sector as a means of accelerating capital accumulation for rural and urban development and to provide manufactured inputs for agriculture and for helping to meet the essential material needs of their entire populations at the lowest possible cost. Simultaneously, they will have to grapple with providing meaningful work for their non-farm rural population.

For industrialized societies, the principal task will be to plan for the rational introduction of this technological revolution so that the extraordinary increases in productivity, which it appears to promise, will be equitably shared by all sectors of society and will facilitate an orderly transition to renewable energy and raw material sources. For these countries, as for developing countries, a major challenge will be to devise meaningful patterns of work for those displaced from primary productive activity by these technological changes.

For the Third World, the micro-electronic revolution also offers the promise, however remote, of a more autonomous, less dependent role in the global political economy. Social equity within developing countries may be no greater, but it may also be no worse than it is now. And for Northern countries, there is the possibility of a somewhat more distant but less exploitative relationship with the South.

But for both groups of countries, the critical problems are economic and political — in a word, who controls, who benefits, and who pays. It would be naive in the extreme to think that the distributional problems generated by these technological changes will be easily or equitably solved as long as control over their development and use is concentrated in a handful of powerful corporate or other dominant political and economic institutions.

Since there is no practical way of getting the micro-electronic genie back in the bottle, we had best get it all the way out. These new technologies will be less likely to be used to perpetuate exploitation and inequality to the extent access to their generation and application is made as open as possible on a global scale. That should be the goal of any effort by the international community to direct and shape this technological revolution.

A key step in devising strategies to achieve this goal will be a critical examination of the relationship between economic and political power and control of revolutionary technologies such as micro-electronics. And that should be the next chapter to be written by those who claim to be students of the changing political economy of the modern world.

Notes

1. See, for example, Barry Commoner, "The Solar Transition," *The New Yorker,* April 23, 1979, pp. 53–98, and April 30, 1979, p. 46–93, and the various writings of Amory Lovins on "Soft Energy Paths," many of whose ideas are pulled together and summarized in a recent article, "Safe Energy," *Resurgence,* January–February, 1979. pp. 17–27. For a series of vignettes of what is just over the horizon in the realm of materials technology, see a series of articles on the theme, "The Future of the Chemical Industry," in *The Chemical Engineer,* June 1979.

2. See, for example, No. 13 (June 1979) of the *Lund Letter on Science, Technology, and Basic Needs,* pointing out that preparations for the UN Conference on Science and Technology for Development (held two months later) had up to then almost completely neglected this issue (and little attention was given at the Conference itself).

3. E.g., P. H. Abelson and A. L. Hammond, "The Electronic Revolution," *Science,* March 18, 1977, pp. 1089–91; A. Burkitt, "Microprocessor Revolution Could be Greater than Them All," *Engineer,* April 27, 1978, p. 39; G. Bylinksky, "Here Comes the Second Computer Revolution," *Fortune,* November 1975, pp. 134–138, 182–184; T. Forester, "The Micro-electronic Revolution," *New Society,* November 9, 1978, pp. 330–333, as cited in Keith Dickson and John Marsh, *The Microelectronics Revolution: A Brief Assessment of the Industrial Impact with a Selected Bibliography,* Birmingham: Technology Policy Unit, University of Aston, *Occasional Paper,* December 1978.

4. Presentation by J. R. Bessant, Technology Policy Unit, University of Aston at Birmingham, to the Lund Workshop on Technological Change in Industrialized Countries and Its Consequences for Developing and Industrialized Countries, Lund, Sweden, May 28–30, 1979 (hereinafter cited as the Lund Workshop).

5. This list is based upon various presentations at the Lund Workshop, especially those by Ruphael Kaplinsky and John Bessant.

6. Presentation by Bernard Real, Institut de Recherche Economique et de Planification, Universite des Sciences Sociales de Grenoble, to the Lund Workshop. Real is engaged in an extensive survey of technological changes in the machine tool industry for OECD. See also his Lund Workshop paper, "Technological Change and Its Economic Consequences: The Case of Machine-Tool Industry" (May 1979).

7. Ray Curnow, "The Future with Microelectronics" *New Scientist,* April 26, 1979, p. 254. This article draws upon a recently published study by Curnow and Jann Barron, *The Future with Microelectronics,* London: Frances Pinter Ltd., 1979. For Porat's categorization of the U.S. work force has been criticized because of his overly inclusive definition of the "Information Sector," but it is still useful for illustrating broad historical trends and proportions as long as the percentages indicated are not used as precise measurements.

8. *Ibid.* J. Herbert Holloman, former U.S. Assistant Secretary of Commerce for Science and Technology and now Director of the Center for Policy Alternatives at the Massachusetts Institute of Technology (which has done several studies of the industrial impact of automation), has used a similar phrase in recent speeches.

9. P. G. Bishop, W. S. Jones, and A. V. Wells, "Microcomputers: Miracle or Myth?," *Electronics Power,* March 1978, pp. 196–200, as cited in Dickson and Marsh, *op. cit.*

10. Staffan Jacobsson, "Technical Change in the Industrialized World and Effects on Underdeveloped Countries — The Aspects of Employment and Technological Dependence (Background paper for Lund Workshop, May 1979). See also his paper, "Technical Change, Employment and Distribution in LDC's" Lund: Research Policy Institute, June 1979, which deals with a number of issues explored in the present paper.

11. The papers for the Bon Worskhop on Technological Dependence are being edited for publication by the organizer of the Workshop, Dieter Ernst, Projekt Technologietransfer, University of Hamburg, Federal Republic of Germany, and will be published in Spring 1980 under the title, *The New International Division of Labour, Technology and Underdevelopment: The Consequences for the Third World,* Frankfurt/Main: Campus Verlag. Micro-electronic technology also appears to have the potential of being capital saving, which would in theory help to balance better the factor proportions in developing countries, but at the present time this remains more theoretical than actual.

12. Institut für Systemtechnik unter Innovationforschung, *Der Einfluss neuer Techniken auf die Arbeitsplätze,* Karlsruhe, 1978, and I. Barron and R. C. Curnow, *The Future of Information Technology,* Science Policy Research Unit, mimeo., 1978 (since published by Frances Pinter — see note 7), as cited in C. Freeman, "Technology and Employment: Long Waves in Technical Change and Economic Development" (Holst Memorial Lecture, December 1978). Some U.S.-based studies suggest a significant impact on employment as does the British study cited above; see for example, Christopher J. Barnett *et al., Industrial Automation — Its Nature, Effects and Management: Critical Issues Report,* Cambridge: Massachusetts Institute of Technology, September 1978.

13. Freeman, *ibid.,* pp. 6–11.

14. *Ibid.,* p. 12.

15. *Ibid.,* p. 14.

16. *Ibid.,* p. 19.

17. Bessant. Presentation to Lund Workshop, *op. cit.;* Dickson and Marsh, *op. cit.* See also Bessant, "An Overview of the Impact of Microelectronics on Manufacturing Industry" (Background paper for Lund Workshop, May 1979).

18. S. Nora, L'Informationisation de la Societie, published by *Documentation Francaise,* May 19, 1978, as cited in Dickson and Marsh, *ibid.*

19. Lovins, *op. cit.;* Commoner, *op. cit.*

20. Wassily Leontief, "Newer and Newer and Newer Technology, with Little Unemployment," *New York Times,* March 6, 1979.

21. The "automation dividend" can, of course, be passed along to the consumer in the form of markedly lower prices, the pocket calculator being the prime example usually offered. Most analysts of the impact of micro-electronics on productive activity, whether in manufacturing or the service sector, however, regard that as an atypical case. Nonetheless, it is almost certain that in most cases there will be some sharing of the "automation dividend" among workers, managers, and consumers. A more critical political question is likely to be who will pay most heavily the social costs involved in automation.

22. It usually contended that Swedish unemployment figures are "artificially" lowered by enrolling workers in government-supported training schemes, public service employment, and subsidized private sector employment. These are likely to be the kinds of measures that will have to be used by industrialized economies in grappling with large-scale technological unemployment in the next decade. Whether palliatives or not, they begin to confront the issue of breaking the link between income and productivity, which will become a paramount political issue in these countries in the coming decade where it is not already.

23. C. Freeman, "Technical Change and Unemployment" (Paper presented to Conference on "Science, Technology and Public Policy: an International Perspective," University of New Wales, December 1–2, 1977), p. 22.

24. For a vigorous, socially conservative attack on the doom criers of two decades ago, see Yale Brozen, "Automation and Jobs," Chicago: Graduate School of Business, University of Chicago, Selected Papers No. 18, 1965.

25. European Trade Union Institute, *The Impact of Microelectronics on Employment in Western Europe in the 1980s,* Brussels: ETUI, 1979, pp. 73–74. The contrast between FRG and Sweden is instructive. During the same period (1971–1975), labor productivity in Sweden rose by an average of 2.5 percent a year and GNP, by 1.9 percent. Working time fell by 1.5 percent so that increase in output and decrease in working time more than offset the increase in productivity, as well as the increase in working population. The result was that unemployment fell from 1.5 percent to 1.4 percent during this period. (*Ibid.,* p. 74).

26. *Ibid.,* pp. 92–95.

27. Sandberg, ed., *Trade Unions, Planning and Computers,* Swedish Centre for Working Life, 1978, as cited in ETUI, *op. cit.*

28. ETUI, *op. cit.,* pp. 80–90.

29. Ray Curnow, private communication to the author, December 1979. Curnow also makes the point that micro-electronic based technology saves capital in its substitution role, with the consequence that not just the capital-labor ratio is being changed, but also the technological coefficient in the production function.

30. ITUI, *op. cit.,* pp. 105–106.

31. *Ibid.,* pp. 106–107. The forecast is from Prognos, as quoted in the ILO *Social and Labor Bulletin,* February 1979.

32. Ray Curnow offers the intriguing thought that even conservative governments may, in the 1980s, take steps to flatten the income distribution curve in order to maintain economic activity levels by shoring up consumer demand.

33. The ETUI report cited above includes many such examples of trade union concern with the technology/job issue. A good illustration of the focus on micro-electronics is the report of a one-day conference in Liverpool, "Microprocessors: The Unknown Future" (Sponsored by the Workers' Educational Association and various trade unions at the University of Liverpool, 28 April 1979). In North America, labor concern with the issue is perhaps not yet as vocal but is growing. See Markley Roberts, "Harnessing Technology: The Workers' Stake," *AFL-CIO American Federationist,* April 1979. The issue is also looming larger in collective bargaining in certain industries such as the automotive: see, for example, 'UAW Fears Automation Again,"

Business Week, March 26, 1979, and "New Auto Industry Technology A Source of Wonder and Anxiety," *New York Times,* November 5, 1979.

34. The New Alchemists, a scientifically sophisticated, socially concerned appropriate technology group in the United States, will shortly be publishing a book, *Tomorrow Is Our Permanent Address,* which argues the proposition that "small computers allow us to provide 'memories' for the architectural/solar/biological entities being created by us. Further in a very cursory way we suggest that electronics can be a soft technology integrating normally disconnected functions such as energy production, architecture and food cultivation." (Letter from John Todd, Director, The New Alchemy Institute, to the author, December 1, 1979).

35. James H. Weaver, Kenneth P. Jameson, and Richard N. Blue, "A Critical Analysis of Approaches to Growth and Equity" (Paper prepared for International Studies Association Annual Meeting, March 17-20, 1977, St. Louis).

36. Pranab K. Bardhan, "The Pattern of Income Distribution in India: A Review" *Sankhya: The Indian Journal of Statistics,* Vol. 36, Series C, Parts 2 and 4 (1974), pp. 103–138, and V. M. Dandekar and N. Rath, *Poverty in India,* Poona: Indian School of Political Economy, 1971.

37. National income distribution statistics on China are not, to my knowledge, available. I am indebted to John Sigurdson of the Lund Research Policy Institute, who has done extensive work on China's rural development, for this characterization.

38. Pramit Chaudhuri, Institute for Development Studies, University of Sussex, Presentation to Lund Workshop and background paper for the workshop, "Labour-Absorption Possibilities in Indian Agriculture: A Brief Report" (May 1979). For figures on landless agricultural laborers in India, see K. N. Ray, "The Economic Situation," *Economic and Political Weekly,* July 3, 1976.

39. Raj, *ibid.*

40. Government of India, Ministry of Finance, *Economic Survey, 1976,* Delhi: Division of Publications, 1976. Even when adjustments are made because of major shifts from the private to the public sector during this period (e.g., nationalization of banks, coal mining), and manufacturing employment in the modern public industrial sector is taken into account, the picture does not change materially.

41. Staffen Jacobsson, "Trends in Manufacturing Output and Employment in LDCs and DCs (Background paper prepared for Lund Workshop, May 1979).

42. Jon Sigurdson, intervention at Lund Workshop.

43. Chaudhuri, Presentation and Background Paper, Lund Workshop, *op. cit.*

44. Charles Edquist and Olle Edquist use the phrase "social transformation" in making this point, observing that social transformations do not necessarily have to be violent. Most of the transformations from feudalism to capitalism were largely non-violent. These transformations "made possible the immense technological progress and increase in productivity that was a part of the industrial revolution. . . . For the underdeveloped countries a repetition of the industrial revolution is impossible. The underdeveloped countries have to find altogether different paths." (Edquist and Edquist, *Social Carriers of Technology for Development,* Lund: Research Policy Institute, University of Lund, *Discussion Paper* No. 123, October 1978, p. 84.)

45. Jacobsson, "Technical Change in the Industrialized World," *op. cit.*

46. Raphael Kaplinsky, Institute for Development Studies and Science Policy Research Unit, University of Sussex, Presentation and "The Impact of Micro-electronic Related Innovations on UDC-DC Trade." (Notes prepared for Lund Workshop, May 1979). The problems of buying advanced technology (as opposed to the products of advanced technology) is well illustrated in a recent article about the electronics industry in India, Smit K. Pal and B. Bowonder, "High Technology and Development in India," *Futures: The Journal of Forecasting and Planning,* August 1978, pp. 337–341.

47. Curnow, *op. cit.*

48. Korea is an example and is now seeking to place its engineers in United States industry to gather relevant working experience through industrial internships in electronics. (Interviews by the author with officials of the United States and Korean electronics industry, California, November, 1979). Staffan Jacobsson points out that small industrialized countries like Sweden face a similar problem in capturing the benefits of a technology of which they are not a producer but only a user.

49. Robert S. McNamara, Address to the United Nations Conference on Trade and Development (Manila, May 10, 1979), Washington: World Bank, May 1979.

50. This thesis has been developed by Claude Alvares in his "Development against People," *Development Forum,* July 1978.

51. I am indebted to Staffan Jacobsson for raising this question.

52. Graciela Chichilnisky, "Basic Needs and the North-South Debate: Imperatives and Distortions" (Draft Paper prepared for the World Order Models Project, May 1979).

53. Kaplinsky, Lund Workshop Notes, *op. cit.* These figures are subject to the criticism that they are only planning projections and not realistic targets for production. But they nonetheless illustrate by a general order of magnitude the problem of unrealistically heightened expectations.

54. But not everyone has reached the same conclusion. Fairchild is reported to be expanding its electronics production facilities in Singapore with increasingly more automated technology. (*Financial Times,* November 26, 1979.)

55. Organisation for Economic Cooperation and Development, *The Impact of the Newly Industrializing Countries on Production and Trade in Manufactures: Report by the Secretary-General,* Paris: OECD, 1979.

56. I am indebeted to Raphael Kaplinsky of the Science Policy Research Unit at Sussex for the idea of the possible impact of micro-electronics on shifting comparative advantage. (Kaplinsky, Lund Workshop Notes and Presentation, *op. cit.*)

57. The basic point is Kaplinsky's (*ibid.*), but I would not want to hold him accountable for my particular formulation of the point.

58. Commoner, *op. cit.* Dieter Ernst of the Projekt Technologietransfer, Universität Hamburg, observes that industrialized country R & D on synthetics and other forms of "substitutive research" induced by price increases in raw materials are likely to undercut Third World efforts to use raw materials as a bargaining lever. (Letter to the author, 21 June 1979.)

59. See Sanjaya Lall, "Developing Countries as Exporters of Industrial Technology," in H. Giersch, ed., *International Economic Development and Resource Transfer,* Tübingen: J. C. B. Mohr, 1979; and David Chu and Ward Morehouse, *Third World Cooperation in Science and Technology for Development,* New York: United Nations Institute for Training and Research, *UNITAR Working Paper Series on Science and Technology No 5,* June 1979.

60. This skepticism about the social consequences of technological and other forms of economic cooperation among Third World countries is reflected in a recent paper by Dieter Ernst, "Technical Cooperation among Developing Countries (TCDC) — A Viable Instrument of Collective Self-Reliance?" (Paper prepared by the Symposium on Science and Technology in Development Planning, UN Advisory Committee on the Application of Science and Technology to Development and other UN bodies, El Colegio de Mexico City, May 28–June 1, 1979).

61. Private communication to the author from Jack Baranson, Developing World Industry through Technology, Washington, November 1979.

62. Interviews by the author with officials of the United States semiconductor and Korean electronics industries, California, November 1979. A similar conclusion is reached in a recent study of North-South technology flows in consumer electronics not yet released by a major intergovernmental economic research organization. Pal/Bowonder article cited above on the electronics industry in India makes the same general point.

63. J. B. Donges and James Reidel, "The Expansion of Manufactured Exports in Developing Countries: An Empirical Assessment of Supply and Demand Issues," *Weltwirtschaftliches Archiv: Review of World Economics,* 1977 (Band 113, Heft 1), pp. 58–85.

64. OECD Report on the NICs, *op. cit.*

65. Louis Turner of Chatham House sees "a complex world in which bits of industry shuffle back and forth between North and South depending on the current comparative competitive position of cheap labour and expensive labour-with-automation." (Letter to the author, July 16, 1979). Juergen B. Donges makes a similar case by pointing out the considerable capacity for adjustment shown by outward-oriented economies of 15 NICs he and his colleagues have studied extensively. See his "A Comparative Survey of Industrialization Policies in Fifteen Semi-Industrial Countries," *Weltwirtschaftliches: Review of World Economics,* 1976 (Band 112, Heft 4), pp. 626–659.

66. See, e.g., the OECD report on the NICs, *op. cit.*

Index

About the Authors

Victor Basiuk is a consultant in science and technology policy and national security policy in Washington, D.C. He taught at Columbia University, Naval War College, and Case Western Reserve University, and was for several years Research Associate at Columbia's Institute of War and Peace Studies. In Washington, Dr. Basiuk served as adviser to the Chief of Naval Operations and was consultant to the Executive Office of the President, Office of the Under Secretary of Defense for Research and Engineering, and several U.S. government agencies. He is the author of *Technology, World Politics and American Policy* (Columbia University Press, 1977), a monograph on *Technology and World Power* (Foreign Policy Association, 1970), and numerous articles and contributions to collective volumes.

Fabio Erber is on the faculty of the Centro de Ciencias Juridicas e Economicas, Universidado Federal do Rio de Janeiro. He received his PhD from the University of Sussex and has been affiliated with its Science Policy Research Unit. He has also been associated with FINEP in Brazil. He has published widely on questions of technology and development.

Robert G. Gilpin is a Professor of Politics and International Affairs, Eisenhowever Professor of International Affairs, Faculty Associate of the

Center of International Studies at Princeton University. He has also served as a consultant to Stanford Research Institute, Ford Foundation Program on International Economic Order, President's Office of Science and Technology, and the Organization for Economic Cooperation and Development. He is the author of many books and numerous articles and monographs.

H. Jeffrey Leonard is a doctoral candidate in the Department of Politics, Princeton University. He received his BA from Harvard University and his MSc from the London School of Economics and Political Science. He has been affiliated with the Conservation Foundation since 1976 and has published several articles dealing with technological environmental politics in the US and abroad.

John D. Montgomery is a Professor of Public Administration, the Kennedy School of Government, and Chairman, Department of Government at Harvard. His most recent books are *Patterns of Policy* (1979) and *Technology and Civic Life* (1974). He has served as consultant to the Department of State, the Agency for International Development, the Department of Health, Education, and Welfare, and the Department of Agriculture, as well as to the governments of Iran, Malaysia, Turkey, Zambia, Colombia, Korea, the Philippines, and others.

Ward Morehouse is president of the Council on International and Public Affairs, research associate at the Columbia University School of International Affairs, and a Fellow of the Center for International Policy in Washington. He was a visiting professor at the University of Lund in Sweden in 1976–77. Morehouse has been a Senior Fellow at the East-West Center in Honolulu, visiting professor at the Administration Staff College in India, and consultant to the Ford Foundation. He was for more than a decade director of the Center for International Programs and Comparative Studies of the University of the State of New York. He is the author and editor of numerous works, including *American Labor in Changing World Economy* (New York: Praeger Publishers, 1978), *Science and the Human Condition in India and Pakistan* (New York: Rockefeller University Press, 1968), and *Science, Technology, and the Global Equity Crisis: New Directions for United States Policy* (Muscatine, Iowa: Stanley Foundation, 1978).

Jurgen Schmandt is professor of public affairs, University of Texas at Austin. His publications are in the areas of science policy, social policy, and policy analysis. Professor Schmandt received his doctorate in Political Philosophy from the University of Bonn, worked with public agencies in Germany, France, and the United States, and spent five years at Harvard University.

Frances Stewart is a Fellow of Somerville College and a Senior Research Officer at the Institute of Commonwealth Studies, Oxford. She has written widely on questions of Third World development, and especially on technology and employment questions. Publications include *Technology and Underdevelopment,* Macmillan (& Westview), 1977. She has been a consultant to a number of institutions including the World Bank, the Ford Foundation, UNCTAD, UNIDO the ILO, and the National Science Foundation.

Joseph S. Szyliowicz is a professor at the Graduate School of International Studies, University of Denver, and Director of its Technology, Modernization, and International Studies Program. He has published widely on questions of technology and change, especially in regard to the Middle East. Recent publications include *Education and Modernization in the Middle East* (Cornell University Press, 1973), *The Energy Crisis and US Foreign Policy* (Praeger, 1975), co-author and co-editor, and *Decision Making in a Technological Environment: The Case of the Aswan High Dam* (International Case Clearing House, 1980), co-author. He has served as a consultant to UNESCO. NSF, HEW, and the Colombo Plan.

Christopher Wright, Staff Member for Science Policy and Institutional Development at the Carnegie Institution of Washington, is the co-author and editor of *Scientists and National Policy Making* (with Robert Gilpin, Columbia, 1964) and *Science Policies of Industrial Nations* (with T. Dixon Long, Praeger, 1975). He was Director of the Institute for the Study of Science in Human Affairs at Columbia University and a member of the faculty from 1966 to 1972. Subsequently he was on the staffs of the Rockefeller Foundation and the Congressional Office of Technology Assessment before joining the Carnegie Institution in 1979.